Advances in Intelligent Systems and Computing

Volume 261

Series editor

Janusz Kacprzyk, Polish Academy of Sciences, Warsaw, Poland
e-mail: kacprzyk@ibspan.waw.pl

For further volumes:
http://www.springer.com/series/11156

About this Series

The series "Advances in Intelligent Systems and Computing" contains publications on theory, applications, and design methods of Intelligent Systems and Intelligent Computing. Virtually all disciplines such as engineering, natural sciences, computer and information science, ICT, economics, business, e-commerce, environment, healthcare, life science are covered. The list of topics spans all the areas of modern intelligent systems and computing.

The publications within "Advances in Intelligent Systems and Computing" are primarily textbooks and proceedings of important conferences, symposia and congresses. They cover significant recent developments in the field, both of a foundational and applicable character. An important characteristic feature of the series is the short publication time and world-wide distribution. This permits a rapid and broad dissemination of research results.

Advisory Board

Chairman

Nikhil R. Pal, Indian Statistical Institute, Kolkata, India
e-mail: nikhil@isical.ac.in

Members

Emilio S. Corchado, University of Salamanca, Salamanca, Spain
e-mail: escorchado@usal.es

Hani Hagras, University of Essex, Colchester, UK
e-mail: hani@essex.ac.uk

László T. Kóczy, Széchenyi István University, Győr, Hungary
e-mail: koczy@sze.hu

Vladik Kreinovich, University of Texas at El Paso, El Paso, USA
e-mail: vladik@utep.edu

Chin-Teng Lin, National Chiao Tung University, Hsinchu, Taiwan
e-mail: ctlin@mail.nctu.edu.tw

Jie Lu, University of Technology, Sydney, Australia
e-mail: Jie.Lu@uts.edu.au

Patricia Melin, Tijuana Institute of Technology, Tijuana, Mexico
e-mail: epmelin@hafsamx.org

Nadia Nedjah, State University of Rio de Janeiro, Rio de Janeiro, Brazil
e-mail: nadia@eng.uerj.br

Ngoc Thanh Nguyen, Wroclaw University of Technology, Wroclaw, Poland
e-mail: Ngoc-Thanh.Nguyen@pwr.edu.pl

Jun Wang, The Chinese University of Hong Kong, Shatin, Hong Kong
e-mail: jwang@mae.cuhk.edu.hk

Pierre-Jean Benghozi · Daniel Krob
Antoine Lonjon · Hervé Panetto
Editors

Digital Enterprise
Design & Management

Proceedings of the Second International
Conference on Digital Enterprise Design
and Management DED&M 2014

 Springer

Editors
Pierre-Jean Benghozi
CRG
Ecole Polytechnique
Palaiseau Cedex
France

Daniel Krob
LIX / DIX
Ecole Polytechnique
Palaiseau Cedex
France

Antoine Lonjon
MEGA International
Paris
France

Hervé Panetto
University of Lorraine
Vandoeuvre-les-Nancy Cedex
France

ISSN 2194-5357   ISSN 2194-5365   (electronic)
ISBN 978-3-319-04312-8   ISBN 978-3-319-04313-5   (eBook)
DOI 10.1007/978-3-319-04313-5
Springer Cham Heidelberg New York Dordrecht London

Library of Congress Control Number: 2013957361

Printed on acid-free paper

Springer is part of Springer Science+Business Media (www.springer.com)

Welcome from Françoise Mercadal-Delasalles, Head of Corporate Resources and Innovation Division, Société Générale

Société Générale is proud to welcome the second edition of the DED&M conference as a partner of this event.

This year the Société Générale Group celebrates its 150th anniversary.

150 years of development, international expansion, diversification but also many years of crises, turbulence and intense mutations that the company and its environment have been touched and changed by.

As many other large corporations, over many decades, we have demonstrated the strength and resilience of our enterprise culture.

But we also realize that the challenge of the digital age is the dawn of a new era and an unprecedented shift in our future that needs to drive us towards reinvention.

To tackle this change of age, we, as companies, need to simplify our different approaches, open ourselves up to the outside, create innovative ecosystems and form new partnerships.

All of the links with the academic world are particularly useful and fruitful in order to feed our ideas about digital transformation and leave behind our organizational complexity to design and manage the digital enterprise of tomorrow. I hope this conference will give us the opportunity to come together to share promising insights about this digital challenge that inspires us all.

Preface

Introduction

This volume contains the proceedings of the Second International Conference on « Digital Enterprise Design & Management » (DED&M 2014 ; see the conference website for more details: http://www.dedm2014.dedm.fr/).

The DED&M 2014 conference was jointly organized by the Dassault Aviation – DCNS – DGA – Thales – Ecole Polytechnique – ENSTA ParisTech – Télécom ParisTech "Engineering of Complex Systems" chair, the Orange – Ecole Polytechnique – Télécom ParisTech "Innovation & Regulation" chair and the C.E.S.A.M.E.S. (Center of Excellence on Systems Architecture, Management, Economy and Strategy) non-profit organization from February 4 to 5, 2014 at the Société Générale in Paris (France).

The conference benefited of the permanent support of many academic organizations such as CEISAR, Ecole Centrale de Paris, Ecole Polytechnique, Ecole Supérieure d'Electricité (Supélec) and Télécom ParisTech, which were deeply involved in its organization.

Our special thanks go to Société Générale which hosted freely the DED&M 2014 conference. We also thank Air France, Orange and MEGA International companies which were the main professional sponsors of the conference. All these institutions contributed to the success of the conference.

Why a DED&M Conference?

Digital enterprises are emerging, but real transformations that will bring digital concepts, uses and processes at the heart of organizations and of their business models are still to come. There is a real stake, on one hand for companies that must understand this evolution and appropriate it as a genuine new enterprise paradigm and on the other hand, for the academic world to develop suitable research activities and adapted skills. That is why mastering digital systems requires an integrated understanding of professional practices as well as sophisticated theoretical techniques and tools.

To do so, we believe that it is crucial to create an annual *go-between* forum at international level, opened to all academic researchers and professional practitioners who are interested in the design and the governance of digital systems from an Enterprise Architecture perspective. The "Digital Enterprise Design & Management" (DED&M) conference meets exactly this objective. It aims to become a key place for international presentations, debates, meetings and exchanges on the Enterprise Architecture dimension of the digital transformation. For its second edition, our event intended again to put digital issues at the heart of its program, but also to bring together business and technological stakeholders of the Digital Enterprise.

The DED&M conference scope consequently integrates both the digital customer & business dimensions (new digital customers' behaviors, digital strategies, proposal and distribution of digital value, digital marketing, digital resources management and governance, digital corporate partnerships, etc.) and the underlying technological dimension (information & communication technology, information systems architecture, database & software engineering, systems and networks engineering, etc.).

The DED&M Academic-Professional Integrated Dimension

To make the DED&M conference this convergence point of the digital enterprise academic and professional communities, we based our organization on a principle of parity between academia and business (see the conference organization sections in the next pages). This core principle was implemented as follows:

- the different Organizing, Program and Strategic Committees of the conference are formed equally both of academic and professional members,

- the Invited Speakers come both from academic and professional environments.

The set of activities of the DED&M 2014 followed the same principle which leads us to propose a mix of research seminars & experience sharing and academic articles & professional presentations during the conference. The DED&M topics cover in the same way the most recent trends in the field of digital enterprise fundamentals and practices from a professional and an academic point of view, including the main professional domains and scientific & technical areas.

The DED&M 2014 Edition

For this second DED&M edition, 34 papers were submitted and the conference Program Committee selected among them 9 regular papers to be published in the conference proceedings. Only the best papers were selected in order to guarantee the high quality of the presentations. 10 complementary abstracts, presented as posters during the conference, were also integrated in the proceedings in a specific dedicated part.

Each submission was assigned to at least two Program Committee members who carefully reviewed it (in many cases with the help of external referees). These reviews were discussed by the Program Committee during a physical meeting held in C.E.S.A.M.E.S. in October 2013 and via the EasyChair Conference management system.

A special focus was put this year on how digital transformation can lead to the emergence of new digital business models. We choose 8 outstanding speakers with various professional and scientific expertises who gave a series of invited talks on this topic. The first day was in particular dedicated to these 8 high-profile invited presentations in order to give to the participants a clear, synthetic and large vision of the domain. An open discussion followed by posters presentations completed this first day that ended into the conference dinner in "Les Ateliers de Renault" on the Champs Elysées in the very center of Paris. The second and last day of the conference was devoted to all contributed talks as well as to two Enterprise Architecture tutorials. "Best papers awards" were announced at the end of the day by the Program Committee chairmen as well as by the president of C.E.S.A.M.E.S. A farewell cocktail finally ended the conference.

Acknowledgements

We would like finally to thank all members of the Program, Organizing and Strategic Committees for their time, effort and contributions to make DED&M 2014 a top quality conference. A special thank is addressed to the team of the Center of Excellence on Systems Architecture, Management, Economy & Strategy (C.E.S.A.M.E.S. ; website: http://www.cesames.net/en), the non-profit organization which managed permanently with an huge efficiency all administration, logistics and communication of the DED&M 2014 conference.

The organizers of the conference are also greatly grateful to the following sponsors and partners without whom the DED&M 2014 would not exist:

Academic Sponsors

- Ecole Polytechnique
- ENSTA ParisTech
- Télécom ParisTech

Professional Sponsors

- Air France
- BiZZdesign
- Dassault Aviation

- DCNS
- DGA
- MEGA International
- OBEO
- Orange
- Société Générale
- Thales

Institutional Partners

- Digiteo Labs
- Ministère de l'Enseignement Supérieur et de la Recherche
- Ministère du redressement productif, des petites et moyennes entreprises, de l'innovation et de l'économie numérique

Supporting Partners

- AIM
- Cap Digital
- CEISAR
- CNAM
- Ecole Centrale de Paris
- IRT-SystemX
- Ecole Supérieure d'Electricité (Supélec)
- Systematic
- The Open Group
- Université de Lorraine

Paris, February 2014   Pierre-Jean Benghozi – Ecole Polytechnique
                                  Daniel Krob – Ecole Polytechnique & C.E.S.A.M.E.S.
                                        Antoine Lonjon – MEGA International
                                      Hervé Panetto – Université de Lorraine

Conference Organization

Conference Chairs

General Chair

Daniel Krob, institute professor   Ecole Polytechnique, France

Organizing Committee Chair

Pierre-Jean Benghozi, professor   Ecole Polytechnique France

Program Committee Chairs

Hervé Panetto   University of Lorraine, France
(academic co-chair)
Antoine Lonjon   MEGA International, France
(professional co-chair)

Program Committee

The Program Committee consists of 28 members (academic and professional): all are personalities of high international visibility. Their expertise spectrum covers all the conference topics. Its members are in charge of rating the submissions and selecting the best of them for the conference.

Members

Co-Chairs

Hervé Panetto   University of Lorraine, France
(academic co-chair)
Antoine Lonjon   MEGA International, France
(professional co-chair)

Members

Denis Attal	Thales, France
Ian Bailey	Model Futures, Great-Britain
Marc Bidan	Association Information & Management, France
Brian H. Cameron, Ph.D.	Pennsylvania State University, USA
Michele Dassisti	Politecnico di Bari, Italy
Dominique Ernadote	EADS CASSIDIAN, France
Ulrich Frank	Unversität Duisburg, Essen, Germany
Ronald Giachetti	The Naval Postgraduate School (NPS) in Monterey, California, USA
Ricardo Goncalves	New University of Lisbon, Portugal
Ted Goranson	Earl Research and Sirius-Beta, USA
John Gotze	EA Fellows, Danemark
Matthew Hause	Atego, USA
Wolfram Kleis	SAP AG, Germany
Jean-Yves Lignier	Club MOA, France
Florian Matthes	TUM, Germany
Juan-Carlos Mendez	AdN International, Mexico
Eric Monnoyer	ERM Consultant, France
Etienne Noiret	IBM
Ovidiu Noran	Griffith University, Australia
Angel Ortiz Bas	Universidad Politécnica de Valencia, Spain
Hervé Pacault	ORANGE, France
Yves Pigneur	Université de Lausanne, Switzerland
Colette Roland	Université Paris 1, France
Frantz Rowe	Université de Nantes, France
Dinesh Ujoodah	Société Générale, France
Lawrence Whitman	Wichita State University, USA

Organizing Committee

The Organizing Committee consists of 8 members (academic and professional) in charge of the logistical organization of the Conference.

Chair

Pierre-Jean Benghozi, professor	Ecole Polytechnique, France

Members

Christophe Longepé	BNP Paribas, France
Jean-Luc Lucas	Ex Orange / France Telecom, France

Jean-René Lyon	MphasiS Wyde & CEISAR, France
Jacques Printz	CNAM, France
Michalis Vazirgiannis	École Polytechnique, France
Alain Vallee	Télécom ParisTech, France
Sylvie Vignes	Télécom ParisTech, France

Strategic Committee

The Strategic Committee helps to define the strategic orientations of the conference. All its members are coming from top executive management of worldwide leading organizations.

Chair

| Jean-Christophe Lalanne | Deputy CIO, Air France, France |

Members

Marc Alinat	CEO Yunano, SVP Atos International, France
Jean-Max Arbez	Boost, Switzerland
François Bourdoncle	EXALEAD, France
Nicolas Colin	Inspection Generale des Finances, France
Daniel Dardailler	W3C, France
Marko Erman	Thales, France
Hervé Gouëzel	BNP Paribas, France
Françoise Mercadal-Delasalles	Société Générale, France
Grégoire Postel-Vinay	DGSIC, France
Elizabeth Pugeat	Crédit Agricole, France
Thierry Souche	Orange, France

Conference Organization

Invited Speakers

Transformation of Existing Digital Models

Technology : Nils Fonstad , e-Lab director, INSEAD ex-MIT - United States
Medi : Olivier Abecassis, General Director, e-business TF1 - France
Communication: Jullien Ampollini, General Director Digital & Marketing Business, PAGESJAUNES - France
Finance : Xavier Terrasse, Digital Working Program Director, BNP Paribas – France

Emergence of New Digital Models

Pharmaceutical : Sabine Safi and Cedric O'Neill, co-founders, 1001 Pharmacies - France
E-Commerce : William El Kaim, COO, Carlson Wagonlit - France
Information Technologies : Marc Alinat, SVP, Atos International & CEO, Yunano - France
Software Technologies for Business Information Systems : Florian Matthes, Professor, TUM – Germany

Contents

11 The Impact of 3D Printing Technologies on Business Model Innovation 119

Thierry Rayna, Ludmila Striukova

Section B: Posters

12 Accelerating Innovation through Modular Design (API) 135

Nicolas Bry, Richard Hababou

13 The Logic of the Reference in the IT Economy 137

Francis Jacq

14 Conceptual Design and Simulation of an Automotive Body Shop Assembly Line 139

Remiel Feno, Aline Cauvin, Alain Ferrarini

15 Using Models for Building Strong Organizations 141

Bas van Gils

16 Quantifying Risk of Acquisition Portfolios 143

Hassan J. Bukhari, Ricardo Valerdi, Daniel Ward

Section A
Regular Papers

Designing Future Enterprises

Milan Guenther and Dennis Middeke

Abstract. The challenges of today's ubiquitous and hyperconnected digital ecosystems make a clear case for a holistic approach, working towards an enterprise-wide transformation rather than a digital extension. This requires overcoming the silos of the past -- enterprise architecture and information systems, digital marketing and communications, and digitally supported services and operations. Only by making those elements part of one strategy is a business able to transform its enterprise ecosystem in order to deliver on its promise to users. A design approach with appreciation of the dynamics and stakeholder complexity in modern enterprise ecosystems is a suitable foundation to formulate and illustrate such a strategy, drawing on an intensive dialogue with strategic decision-makers. Our Enterprise Design Framework provides practitioners to a map of 20 aspects relevant to such work, looking at a large variety of concerns related to enterprise-people relationships. Developed with various clients, we applied it recently in a project with a United Nations agency, aiming at reshaping their digital strategy.

1   The Relationship Challenge

Enterprises are everywhere, playing a vital role in our lives. Basically all human endeavours have reached a level of scale and complexity that makes them depend

Milan Guenther
eda.c gmbh & co kg
enterprise design associates. consultancy
c/o le laptop
6 rue arthur rozier
75019 Paris
France

Dennis Middeke
eda.c gmbh & co kg
enterprise design associates. consultancy
wilhelm-tell-straße 25
40219 Düsseldorf
Germany
e-mail: hello@eda-c.com

P.-J. Benghozi et al. (eds.), *Digital Enterprise Design & Management*,
Advances in Intelligent Systems and Computing 261,
DOI: 10.1007/978-3-319-04313-5_1, © Springer International Publishing Switzerland 2014

on an ecosystem of interrelated organizations and technology. Most of us rely upon enterprises several times a day, even for the most fundamental tasks of daily life. They are ubiquitous in our world of consumer brands and services, and visible in the mass of organizations of all sizes we are in touch with as consumers, employees, investors, or other roles.

When dealing with enterprises, we are used to strange and sometimes quite frustrating experiences. They seem to make even simple transactions awkward and complex. Straightforward activities such as booking tickets for a journey, paying your taxes, subscribing to health insurance, or resolving a problem with your energy supplier require customers to embark on a laborious journey, jumping between call centres, online forms, apps and systems. They make us shift between different contacts, tools, and communication channels (Rosati, Resmini 2011). They lose track of the conversation, get stuck in inflexible procedures, communicate in bits of incomplete information, and often ultimately fail to deliver what they promised. Such experiences happen with companies, government institutions, or other types of enterprises, making them appear slow, rude, and inhumane. When looking at the big picture of human-enterprise interaction, there are numerous examples of failed relationships. While most of them are simply annoying and just make you go somewhere else, some examples of failed relationships have a profound impact on people's lives. They result in lost customers, demotivated employees, or even scandals being echoed in mass media.

With all the technology enterprises are already employing today, it seems be clear that just introducing more systems and more data for its own sake does not solve these problems. Organisations are seldom leveraging technology to gain competitive advantage, but are merely using digital tools to facilitate and streamline their existing activities. In order to take on the evolved role of information systems that support people in various contexts and facilitate enterprise-people relationships, enterprises need to redesign and restructure themselves around human needs, and appreciate the complexity of human experiences in a digital age.

The design disciplines are traditionally occupied with that exact challenge - designing products, websites, brands or even services based on a deep understanding of human needs, usages and behaviours, combining this with business goals, technical possibilities and creative cultural production. As labels such as Webdesign, User Experience Design, Service or Information Design suggest, these approaches are bound to specific domains and artefacts. In a business context, they are usually employed in isolation, following a logic of execution. In our design practice with large organisations, we looked for ways to bridge the gaps and design in a more holistic fashion. Going beyond Design Thinking, we propose an approach that takes design practice to the enterprise level.

2   The Enterprise Design Idea

The challenges of today's ubiquitous and hyper-connected digital ecosystems make a clear case for such an approach, working towards an enterprise-wide transformation rather than a digital extension. Digital is everywhere, inseparable even from physical interactions and personal services. This ubiquitous, fluid digital layer is in sharp contrast to the way many organisations treat their digital touchpoints with people. A digital strategy is not about having the Marketing department produce websites and Facebook pages, the R&D people work on apps and digital products, field services enhance physical stores, or IT refurbishing information systems and portals. Appreciating the fact that it is not just the quality of one element but the intersections and transitions between those elements that make or break customer experiences (or employee, shareholder or partner experiences for that matter), such an approach would look at everything relevant, attempting to understand and reshape the entire system. This requires overcoming the silos of the past -- enterprise architecture and information systems, digital marketing and communications, and digitally supported services and operations. Only by making those elements part of one strategy is a business able to transform its enterprise ecosystem in order to deliver on its promise to people (Guenther 2012).

The complex and volatile nature of such systems quickly becomes overwhelming, with business processes, enterprise actors, brand interactions, culture, content, business models, technology or touchpoints being just tiny parts of the puzzle. Too often in a classic decision-centric management setting, long lists of requirements as the basis for all further endeavours seem to just magically appear out of nothing and remain unquestioned, instead of being part of a larger vision and purposeful design of the business. When design competency is finally called in, it is too late - the inconsistencies and missing links that turn out in bad experiences are already hard-wired and constrain the solution space (Borja de Mozota 1990).

To avoid this kind of problem, the idea of the enterprise as subject to design work takes a different approach to design: that the key challenges companies and other organizations face are best tackled by addressing them in a strategic design initiative, working in a holistic and coherent fashion. Instead of focusing on an isolated function/department, domain or touchpoint, the design approach seeks to appreciate the complexity of stakeholders and their enterprise-people relationships.

In this context, an enterprise can be seen as a purposeful endeavour, an idea shared by the various people involved, and a set of identities, architectures and experiences to be designed. As the larger playing field for a design initiative, it is both subject to applied design work (doing research, conceptualisation, idea generation, validation) and the context wherein the outcomes are to be applied.

In such a setting, the role of design shifts from producing isolated solutions to well-defined problems to one of relentless inquiry into a complex system, in order to identify the problems or themes to tackle. Problems or themes to be addressed are ill-defined, making potential solutions underdetermined (Camillus 2008). This environment makes the case for a close integration of strategic thinking and design practice.

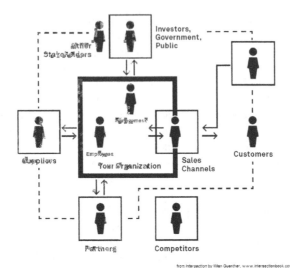

from *Intersection* by Milan Guenther, www.intersectionbook.com

Fig. 1 A typical commercial enterprise ecosystem

3   Strategic Design

When looking at relationships between enterprises and people, it becomes apparent that reshaping and transforming them is no trivial task. Because relationships fail or succeed at the enterprise level, it is also on this level that they need to be considered. In order to address them in a coherent and holistic fashion, design practice must wholeheartedly embrace the complexity and diversity of those relationships. Relationship challenges do not halt at the boundaries of particular disciplines, practices, thinking styles, or knowledge areas. Due to the diversity of domains touched by such a challenge, it is often unclear what kind of competence would be best suited to approach the problem, and who to involve as an expert.

In order to achieve innovation at an enterprise level, design practice needs to look at opportunities for innovation at the intersection of different domains, functions, one disciplines and practices. Instead of centering on a particular viewpoint, practitioners seek inputs and inspiration, constraints and ideas to generate a meaningful, viable and feasible future state (Gall et al 2010). This "archeology of the future" (Shanks 2011) then becomes a tangible evidence of the desired future way, and guides planning, blueprinting and change processes that have a transformative impact on the enterprise and its ecosystem. To do so, the relationship of design and strategy has to be one of a continuous dialogue. Design has to inform strategy, by providing insights on potential innovations, and generating options to choose from. Strategy has to be the basis for any vision made visible by design, from initial visual sketches that capture a future state to the redefinition of the enterprise as a modular system. The enterprise becomes a place to iteratively develop and pursue a portfolio of strategies informed and conveyed by design visions, fostering local experimentation and applying good practice to the wider organization.

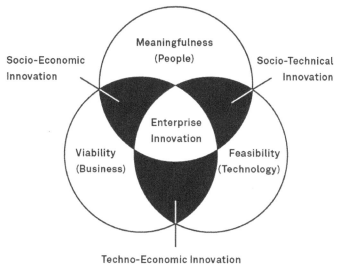

from *Intersection* by Milan Guenther, www.intersectionbook.com

Fig. 2 Innovation in Enterprise Ecosystems

4   Enterprise Aspects

The challenge of designing strategically at the enterprise level makes design work shift, from a traditional focus on a set of isolated aspects (such as online interactions or products) to a comprehensive view of aspects to look at. We developed a set of 20 interrelated aspects loosely corresponding to disciplines and approaches used in strategic design work (Guenther 2012).

It is meant to be used as a map to navigate the complex space of intermingled concerns that is the playing field of strategic design work, providing a common language as well as a checklist of elements to consider. Going from top to bottom in five layers, it allows bridging abstract strategic thinking with conceptual and creative work, aligning disparate elements (Kalbach, Kahn 2011) and turning this into tangible and visible outcomes.

- The **Big Picture** aspects help to understand the enterprise as a whole, being subject to all design activities, and at the same time providing the context for all outcomes. As universal qualities, they apply to any enterprise, even if not consciously addressed. In strategic design work, they enable envisioning potential futures beyond individual stakeholder perspectives.

Fig. 3 The Enterprise Design Framework

o **Identity:** the enterprise as a mesh of personalities, impressions, and images in people's minds, as expressed in symbols, language, and emerging culture. It is subject to Branding work and related initiatives.

o **Architecture:** the enterprise as a purposefully designed system of control structures, managing resources, assets, process flows, and capabilities. It is subject to cross-domain Enterprise Architecture work.

o **Experience**: the enterprise as a space of people, environments, and artefacts. Experience Design work strives to redesign and improve these exchanges, starting from human behaviour and perception.

• The **Anatomy** aspects are all about the loose parts, capturing the volatile, interrelated building blocks that form the enterprise as a dynamic ecosystem. As elements of a fractal structure, they recur across all scales and domains. Applied in research and conceptual design work, these aspects provide the basis to collect, map, understand, co-create and rearrange enterprise elements as part of an intended transformation.

o **Actors:** the variety of stakeholders which are related to the enterprise, addressed or impacted by its activities, or involved in their execution–employees, customers, partners...

- o **Touchpoints:** Whenever a person gets in contact with an enterprise (such as a brand, product, or communication) qualifies as a touchpoint of some sort, with individual journeys from brief encounters to lifelong relationships.
- o **Services:** The concept of services is useful for all kinds of value propositions offered to customers and other stakeholders, made available as the result of enterprise activities (Graves 2009).
- o **Content:** pieces of information or data, which provide meaning to stakeholders. Consistent content is the basis for all communication, decision-making and collaboration.

- Designing at the enterprise level requires working in a complex space of underdetermined problems. Finding a potential solution involves identifying the right questions, often against original assumptions and ideas. The four **Frames** suggest a set of fundamental perspectives to guide conceptual modelling and help deciding on a direction according to strategic choices.
  - o **Business:** This aspect is about market offerings and profitability. Working closely with business stakeholders, it allows expressing in business terms the objectives behind a design initiative. Such an approach focuses on the customer value gained and changes to drive productivity.
  - o **People:** This aspect allows designing in a human and empathetic way, directly working with the people addressed. It is the basis for a Human-Centred Design approach, grounding design decisions in the goals, characteristics, needs, expectations, and individual contexts of real people.
  - o **Function:** The Function aspect captures the purpose the enterprise fulfils and the behaviours it exhibits towards its stakeholders when implementing the outcomes of the design initiative. It supports Requirements Engineering work, eliciting and prioritizing needs together with feasibility experts.
  - o **Structure:** This aspect is about exploring the problem domain in concept models, capturing objects and entities and the way they constitute a larger structure. Collaborating closely with domain experts, it enables a Domain-Driven Design approach (Evans 2008) to transform the structure of the enterprise.

- (Re-)designing an enterprise means actively engaging in change, and achieving coherence across different domains relevant to its endeavours. It involves aligning different viewpoints to unveil opportunities and constraints, ultimately coming to a clear vision of a desired future state. The conceptual aspects forming the **Design Space** provide a map of potential design decisions to be made in order to get there.

- o **Communication:** communication processes, in terms of key messages to be conveyed, and the choice of media to support a social exchange between relevant actors. Such work is driven by the interplay of medium and message, with digital channels used in a physical reality.
- o **Information:** Designing information is about providing the right things to the right people at the right time, drawing out meaning from an ever more quickly growing mass of data. It is subject to Information Architecture work on information use, organization, and classification across the enterprise.
- o **Interaction:** This aspect focuses on connecting people to functions they are using in the enterprise realm, where virtually no activity is carried out without the help of (digital) technology and tools. Interaction Design is about shaping behaviours define and design useful tools and services.
- o **Operation:** The Operation aspect is about the way the enterprise carries out its activities, both human work and automated procedures. Applied in Business Architecture work, it means identifying business drivers and reshaping flows of work to make an enterprise perform.
- o **Organisation:** This aspect is about designing organisational structures to support the enterprise in its activities. This Organisational Design has to take into account the shape of formal roles, incentives and responsibilities, but also their influence on emerging team culture and habits.
- o **Technology:** The Technology aspect is about identifying the technical options for a strategic design challenge, to support human activities and enterprise functions as part of an overarching structure. In design work, this involves a creative elaboration and orchestration of components.

- To deliver on its promise to draw a picture of the future of an enterprise, any strategic design initiative needs to result in visible and tangible outcomes as evidence of an evolved future enterprise. This involves generating ideas going beyond conceptual models, combining the rational with the inspired. A **Rendering** results in a hybrid system of tangible elements that spans the virtual and physical realm of the enterprise. Think of these elements as the triggers for the larger transformation the initiative is aiming for.

- o **Signs:** Systems of signs provide ways for an enterprise to reach out to people, encoding stories and communication flows in media. The various subfields of Media Design give messages and symbols a form in words, pictures or sounds, and allow us to make identity visible, provide interfaces to tools and services, or support wayfinding.
- o **Things:** Things relevant to the enterprise include the objects people use, own, consume, take with them, trade, or create within its realm, both physical and digital artefacts. Designing things is following an Industrial Design approach, selecting materials and characteristics with usage, cultural meaning and marketization in mind.
- o **Places:** Places are where people go, where they live or stay, meet or work. They provide the environment for activities in the enterprise, triggering personal and social associations, memories and moods. Both digital and physical place-making is Architecture work, generating context for people by shaping their surroundings.

Using an adapted industrial design process (Design Council 2006), the Enterprise Design Framework allows making design choices on different levels aligned with strategy. Going through the different phases, different layers of the design framework come into focus. This in turn can be used to realign the various moving parts of enterprise ecosystems in traceable models, making them strive towards a common vision.

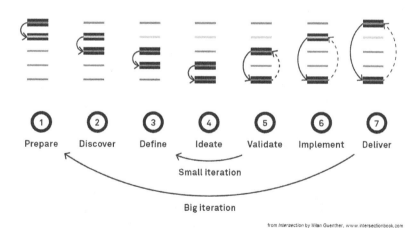

from *Intersection* by Milan Guenther, www.intersectionbook.com

Fig. 4 Applying the Enterprise Design Framework in a typical design process

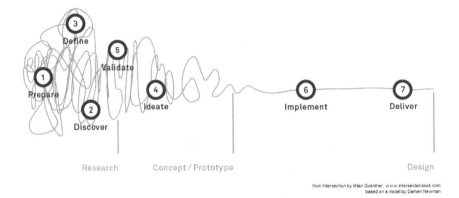

Fig. 5 A more realistic representation of a design process

In reality however, design processes tend to be much messier, difficult to plan and requiring an opportunistic approach to focus areas and applied methodologies. Designers iteratively inquire into the problem space, to make moves in a design space that is constantly being questioned and redefined. While working with a large number of techniques and methods, even for the most methodical designers no single design project is the same. Understanding the problem and inventing a solution both happen in parallel, and ideas for potential outcomes can occur in any phase of the process. One project might make you move slowly from research to definition to ideation, developing a thorough understanding of a complex problem before deciding how to proceed. Another may require developing a first shot in a matter of days and co-creatively reshaping it with peers, looping through the process steps in short cycles.

5   Case Study: The United Nations UNISDR Agency

UNISDR is the UN office for disaster risk reduction, and also the focal point in the UN system for the coordination of disaster risk reduction and the implementation of the international blueprint for this area. Even though the fundamental mission of their enterprise is quite simple - reducing disaster impact and losses worldwide - being an agency within a huge international organisation, UNISDR is dealing with a complex global environment of internal and external stakeholders. We were asked to lead a strategic design initiative to map and explore this enterprise ecosystem, and to design a to-be landscape of digital components that help the organisation to better fulfil its mission. Even though driven by a digital vision, this also included shifts outside the purely digital realm, such as introducing new or changed capabilities or brands.

In workshops, research and co-design we created models of the current, intermediate and future states, navigating the space of intermingled concerns using our Enterprise Design Framework. We started with sketches and lists, then switched to a modelling tool to map and connect the different models. To incorporate aspects relevant to our Enterprise Design challenge, we developed a modified interpretation of the Archimate language (Op 't Land et all 2009), including entities like touchpoints, brands, personas, actors, content, or interactions.

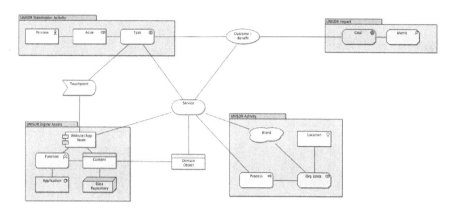

Fig. 6 Relevant enterprise aspects in a metamodel

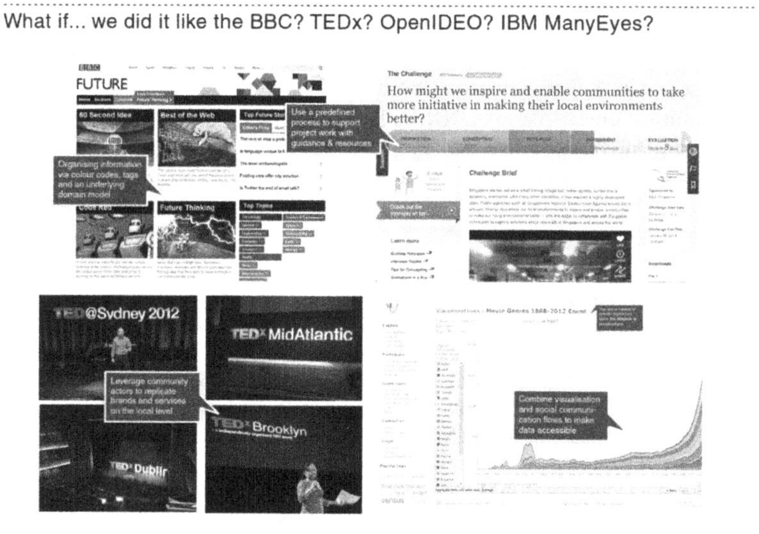

What if... we did it like the BBC? TEDx? OpenIDEO? IBM ManyEyes?

Fig. 7 Outside-in research to inform the creative process

Instead of following a predetermined route, we opportunistically incorporated the aspects that fit our purpose, and adopted the language of our client. Building on views on the enterprise model we generated visualisations of a to-be state, as a basis for senior management decisions and to determine the scope, structures and behaviours of the new digital landscape. This work then directly informed hands-on design work, prototyping key elements and critical moments of the desired experience - expressing intended outcomes in experiential terms, and mapping them to strategic motivations and architectural principles. Tying conceptual design decisions to human-centric research and outside-in benchmarks, we were able to envision "what might be" and turn ideas into innovative solution concepts.

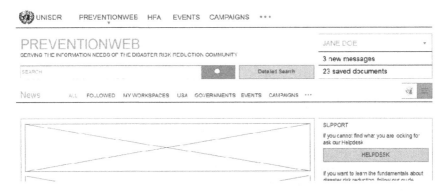

Fig. 8 Wireframe illustrating a component of the future digital platform

This way, we ensured that our design activities were used as inputs to strategic decision-making rather than details of strategy execution. All the digital prototypes and experience scenarios we designed served as tangible evidence of the potential future enterprise of UNISDR, with every creative detail rooted in the blueprint of their business.

References

de Mozota, B.B.: Design Management. Allworth Press, New York (1990)

Camillus, J.C.: Strategy as a Wicked Problem. Harvard Business Review, 98–106 (May 2008)

Evans, E.: Domain-driven Design. Addison-Wesley, Boston (2003)

Gall, N., Newman, D., Allega, P., Lapkin, A., Handler, R.A.: Introducing Hybrid Thinking for Transformation, Innovation and Strategy. Gartner Research (2010)

Graves, T.: The service-oriented enterprise. Tetradian Books, Colchester (2009)

Guenther, M.: Intersection – How Enterprise Design Bridges the Gap Between Business, Technology and People. Morgan Kaufmann Publishers, Boston (2012)

Kalbach, J., Kahn, P.: Locating Value with Alignment Diagrams. Parsons Journal of Information Mapping 3(2), 1–11 (2011)

Land, M.O., Proper, E., Waage, M., Cloo, J., Steguis, C.: Enterprise Architecture. Springer, Berlin (2009)

Resmini, A., Rosati, L.: Pervasive Information Architecture. Morgan Kaufmann, San Francisco (2011)

Shanks, M.: We have always been cyborgs. Archaeology and Things, a Special Issue of Science, Technology & Human Values. SAGE Journals (2011)

Design Council. A Study of the Design Process. The Design Council, London (2006)

The Customer Relationship Management at the Era of Smartphones: Study of the Project Mobile-Dinar within the Arab International Bank of Tunisia

Emna Ben Romdhane and Yosri Bouchioua

Abstract. Our work is interested in understanding the of customer relationship management within the Arab International Bank of Tunisia (BIAT), through the project Mobile-Dinar. The contribution of this work is to understand the factors of failure of the Mobile-Dinar since we noted, following the discussions with senior officers of the BIAT, that this service remains ignored with a very much reduced number of users, and this in spite of its launching in April 2011. To collect information likely to allow us to understand the reasons of failures of this project, we managed a questionnaire with the customers of one of the agency of the bank.

In addition, interviews were carried out with the persons in charge having been implied in this computing project. Recommendations were proposed following the results that we obtained.

Keywords: Customer relationship management, mobile-Dinar, success, failure, Tunisia.

1 Introduction

Smartphones, Facebook, Twitter as well as others are new media of information and communication which the companies could not ignore if it wants to consider the new practices of use and consumption of the customer. This one occupies more than ever the central place in the strategies of the companies and his loyalty becomes a major stake. The customer relationship management (CRM) continues

Emna Ben Romdhane · Yosri Bouchioua
Business School of Tunis
Tunis (Tunisia)
e-mail: {emna.b.romdhane,yosri.bouchioua}@gmail.com

P.-J. Benghozi et al. (eds.), *Digital Enterprise Design & Management*,
Advances in Intelligent Systems and Computing 261,
DOI: 10.1007/978-3-319-04313-5_2, © Springer International Publishing Switzerland 2014

to be the principal vector of this development of consumer loyalty through the integration of the management tools and from now on social networks and smartphones.

The challenge is to offer to the customer the same information and a personalized service; some is the channel which he prefers to use.

The difficulty which arises today is that the companies invest massively in these tools without preliminary reflection on the conditions of their successes [1]. Indeed, these investments always do not lead to a value for the customer because the project of customer relationship management is complex and its strategy must be clearly established up stream [2]. Thus, our work is interested in understanding the factors of success and failure of customer relationship management. We concentrated on the case of the International Arab Bank of Tunisia and particularly the project Mobile-Dinar that the bank developed in 2011to gain the loyalty of its customers but who did not have discounted success.

However, this project is very advantageous for the treatment of the current operations on smartphones and is characterized by the speed of the transfers with weak commissions, the whole in the security and the confidentiality.

It is within this framework that our problems of research are articulated, which arises as follows:

- Why didn't a strong proportion of the customers of the BIAT use the Mobile-Dinar despite all the advantages which are associated to it?
- Which are the factors of failure of this project?

To collect information likely to allow us to understand the reasons of failures of this project, we carried out an exploratory study with an administration of a questionnaire to the customers of the BIAT. In addition, talks were carried out with persons in charge having direct links with the project Mobile-Dinar and its impact on the bank.

This work is articulated around two parts: the first gives a theoretical outline on the customer relationship management and its factors of success and failure. The second presents the management of the customer relation within the bank and is focused on the factors of failure of the project Mobile-Dinar.

2   Theoretical Background

2.1   Factors of Success and Failure of the Customer Relationship Management

If [3] define the customer relationship management as an organizational approach which aims to better knowing and better satisfying the customers, identified by their potential with activity and profitability, through a plurality of channels of contact within the framework of a durable relation in order to increase the turn over and the profitability of the company. Indeed, customer relationship management aims at outwards establishing a continuous and enriched relation with the customer even by the moments by consumption or purchase. This relation is built by the establishment of a dialogue between the company and the customer, being carried out on various channels (mail, telephone, Internet, etc).

It is resulting from a qualitative strategy and tends towards a relation increasingly personalized/individualized with the customer [4].

Before setting up a strategy of CRM, it is essential that the company identifies the key success factors and the factors of failure related to the implementation of such tools. .

2.1.1   The Factors of Success of the Customer Relationship Management

The CRM has as ultimate objective [3] the increase in the turnover and the profitability of the company. With this intention, a careful thought on the factors of success of such an approach is necessary. Several studies were interested in the factors of success of the CRM.

[5] showed in their study on the CRM in the automotive industry that the users and the implementation are the factors of success. The authors explain that success depends on the interest of the employees to accept the change and to the bases of the implementation.

For [6], a clear managerial vision taking into account the user's needs and the material support is essential to the success of the CRM.

[7] were interested in the CRM in the banking environment and emphasized four factors which turn around these tools, namely, the company, staff, technology and the customers.

For their part, [8] concluded that a successful CRM is the combination of information technology (IT), of the support of the direction and the employees.

[9] showed the existence of various factors of success:

- Long-term orientation: as of the first interactions and throughout the relation, the company must express towards the customer his motivation to maintain the exchange. Such an orientation makes it possible to establish a solid bases confidence and shows the sincere commitment on behalf of the company.
- Reciprocity: in a long-term relation, it is not necessary that the partners maximize their benefit in each transaction; the most important is that the total of the relational assessment is balanced. The company must show its customers that it aims at optimizing the mutual profit according to a relation winner /winner;
- Reliability: to show its relational orientation, the company will try to understand the waitings of its customers compared to the tasks to achieve. It will endeavor to fulfill its roles regularly to show its reliability.
- Information exchange: the information exchange which can be useful for the partner represents an undeniable advantage for the two parts and constitutes a proof of confidence.
- Flexibility: in certain situations, it may be that reality does not correspond to the forecasts defined at the time of the agreement. The provision of a supplier to

adapt an agreement to the new conditions of an exchange, for example by modifying the quantities or the delivery periods, expresses its intention to maintain the relation with the customer by respecting the interests of each one.
- Solidarity: in difficult situations, a strong relational orientation can encourage the supplier to propose assistance to the customer (as far as its possibilities but without immediate concrete counterpart.) This help can be more or less material.

2.1.2   The Factors of Failure of the Customer Relationship Management

It can exist barriers to the the application and the success of the CRM. According to [10], the companies invest in tools of CRM without studying the strategic and organizational impacts of these tools on the company.

The absence of these studies does not make it possible the company to evaluate the optimal output of the CRM and to solve the obstacles related to the investment and the implementation.

In their study on the installation of the tools of CRM in the financial sector, [11] concluded that the major brakes with the development of the CRM are the absence of leadership and the quality of the customer's data available to the company.

According to [12] the absence of return on investment of the systems of the CRM is due mainly to the absence of adaptation of the organization to the change in terms of structures and technology.

For its part [13] was interested in the factors of failure related to the implementation of the systems of the CRM in the banking environment and concluded that these factors are: the absence of strategic vision, the absence of implication and the complexity of the structures.

On this subject of factors of failure, [14] retains that:

- The establishment of a personalized and regular communication with the consumers very much: cost of purchase of the database, cost of data processing, and cost of the media (vocal mailing and waiters).
- Much of companies hesitate to invest in the constitution and the exploitation of these databases: large distribution still uses the techniques of mass of marketing.
- Relational marketing exhausts the consumer. It feels tracked, badgered by the telephone, the fax and papers which overflow its letter-box.
- Profitability is not immediate because one does not seek in the short run to generate sales.

3   Operating and Statistical Framework

To retain and satisfy her customers, the Arab International Bank of Tunisia (BIAT) launched, in the month of April 2011, an innovative service in the

Tunisian banking market. Mobile-Dinar service allows the transfer of money using a mobile phone or through a secure website.

3.1   Presentation of the Mobile-Dinar

Within the framework of the widening of the banking services and in accordance with the approach multichannel, making it possible to meet the needs for the customers, the BIAT decided to launch this new service called the Mobile-Dinar.

It is intended to any 18 years old person and more, holder of a mobile telephone line, having or not a bank account.

Thus, any person will be able easily to adhere to the service Mobile-Dinar and to carry out transfers of money with her mobile number. Each customer Mobile-Dinar will have the possibility to transfer instantaneously an amount of money to another customer Mobile-Dinar by indicating simply the phone number of the recipient.

Mobile-dinar is a practical and secure service very indicated for the people having the habit to send/ receive, in a periodic way, transfers of money to their close relations (parents, child, students...) and wishing to avoid displacements at the agency. In some clicks on his mobile phone, the user can indeed carry out transfers of mobile with mobile in full safety. At the moment even where the transfer is carried out, the respective accounts M-Dinar of the transmitter and the receiver are simultaneously debited and credited. An SMS of information on the details of the transfer is displayed instantaneously on their respective telephones.

In short, the Mobile-Dinar changes the vision which one can have on the transfers of money. It is only enough to have two portable devices and to install the M-Dinar application there. On this subject, the BIAT knew to widen the targeted customers by offering two types of applications downloadable directly on its website.

A first application intended for an average range of portable devices equipped with JAVA and which makes it possible to fulfill the same functions as the second application intended for the commonly called intelligent portable "Smartphone". Only one passage to the agency is necessary to authorize the bank to create an account M-Dinar and then everything is done remotely.

This new project seems to be perfect for the users. Nevertheless, it did not succeed in collecting a great popularity within the customers. In what follows, we will see the major causes which slowed down the development of this project and we will propose certain recommendations.

3.2   Methodology

To collect information likely to help us to better understand the factors of failure of the Mobile-Dinar, we carried out an exploratory analysis near the customers of one of the agencies of the bank. Exploratory interviews were also carried out with persons in charge of the bank.

We managed a questionnaire with a convenience sample composed of 150 customers of the BIAT agency, located at Ennasr. The sample is composed of the users and non-users customers of the Mobile-Dinar. The questionnaire was administered face-to-face and the data analysis was carried out using SPSS 19.0.

3.3 Results of the Data Analysis Resulting from the Questionnaire

Our sample is mainly young since 44% have between 25 and 35 years, which let us suppose greater predisposition to the use of information technologies and in particular the adoption of the Mobile-Dinar. Our sample is composed of 68% men and 32% women.

The majority of the questioned people belong to the sector of the services with a rate of 44%, whereas 24% are students and 8% are part of the sector of industry. In addition, 70% of the questioned people use the remote banking services whereas 30% are addressed directly to the agency.

The results show that 22% questioned people know the project whereas 77% do not have any idea of it. Among the 22% of the customers who know the project only a person knew the project through her family, four other people knew the project through their friends, three people knew the project thanks to publicity and three others knew the project through their account manager.

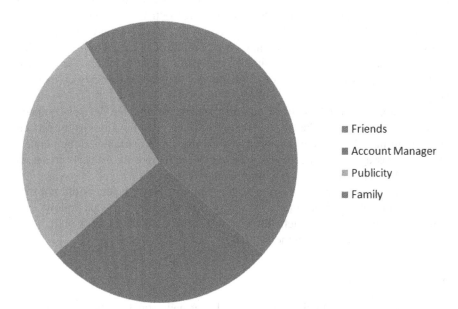

Fig. 1 How the customer heard about Mobile-Dinar

According to the eleven people who know the project, only seven customers took a subscription to this service whereas the four others did not do it. The seven people who use this service find that the M-Dinar is completely secure, easy to use and they are convinced that the project will develop in the future.

Among the seven people who have a subscription M-Dinar, three people suggest having cards like new features and three other people would wish the improvement of the interface. The seven users are very satisfied with the service and three of them find the speed of the transfers as being a major advantage, two respondents indicate the flexibility like major advantage and another person mentions the security.

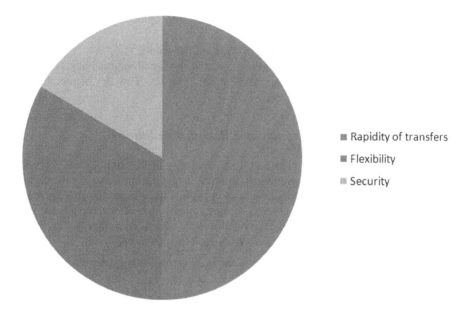

Fig. 2 The Mobile-Dinar Advantages

We noticed that 66.7% of the people who do not have subscription M-Dinar are interested by the project after being informed of the remote services and wished to fill the registration form against 33.3% which was not interested by the M-Dinar.

3.4   Results of the Exploratory Interviews

To be able to determine the main causes of the problems encountered by the Mobile-Dinar, it was necessary to collect information within the head office of the BIAT.

With this intention, we carried out discussions with the designers of this service which taught us that there were only 1350 users for this application.

We understood that the lack of communication is the main factor slowing the project. At its launch, the advertising has been diffused in radios but for a very short period.

Because of the budgetary restrictions, the M-Dinar did not profit from appropriate communication strategy. Additional investments would have made it possible to develop more, especially that the M-Dinar is new on the Tunisian market of the remotely banking services. In front of the rise of the mobile phones and their functions, it seems easy to integrate the project in the practices of use of the potential customers.

The BIAT offered training on this application to account managers by hoping that the agent makes known the M-Dinar, at the time of the exchange with the customer. This proved to be insufficient, taking into consideration the reduced number of users.

A website was set up for this service (www.mdinar.com).

This last explains very well the advantages and the functionalities but the problem, once again, is the lack of communication about it. How the potential customers can be aware of this project?

3.5   Summary of the Results and Recommendations

The Mobile-Dinar is a service ignored by the majority of the sample. We propose to improve the strategy of communication while insisting on the ease of use and the usefulness to support the adoption of such applications in the future. Especially since the majority of non-users was willing to subscribe after having information about the service. By considering the obstacles identified higher, we propose the following improvements that allow M-Dinar to be developed:

- To communicate about this project with the customers by distributing flyers highlighting its advantages;
- To imply, sensitize and train more the account managers of the agency;
- To insert a link in the website of the bank to know the existence of the mobile-Dinar and access the website dedicated to him;
- To decrease the charges on each transfer. Currently the BIAT takes 1% of each transfer of money. Certain current users complain and think about it that it is much.
- To reorganize a good targeted advertising;
- To improve the interface of this application, which is rather basic.

4   Conclusion

The customer relationship management is a strategy which aims at as well as possible, using the resources, competences and the tools available, to understand, anticipate and better meet the needs and expectations of current and potential customers. The company must personalize each relation with its customers in order to offer them services or projects adapted to their needs. It must also have a good communication with its customers, have a good follow up of the orders, invoices and anticipate the market evolution.

The CRM is based on the following principle: it is easier and more profitable to earn the customer's trust and consequently to preserve it longer than to acquire some another, which will be more expensive. It is an instrument of development of the relation of which the goal is to set up a plan to satisfy the customers' needs while betting on the durability of the relation. Nevertheless, there are some principles to respect to avoid the limits which we evoked previously.

References

1. Dickie, J.: Don't confuse implementation with adoption: just because CRM deployments have spread doesn't mean they've succeeded. CRM Magazine (May 2009), http://www.destinationcrm.com/Articles/Columns-Departments/Reality-Check/Dont-Confuse-Implementation-with-Adoption-53684.aspx (accessed on August 2013)
2. Rosset, S., Neumann, E., Eick, U., Vatnik, N.: Customer lifetime value models for decision support. Data Mining Knowledge Discovery 7, 321–339 (2006)
3. Alard, P., Guggémos, P.A.: CRM: Les clés de la réussite. Editions d'Organisation (2005)
4. Lindon, D., Lendrevie, J., Lévy, J.: Mercator, Editions Dunod, 9th edn. (2009)
5. Jomphe, Y., Plaisent, M., Pecquet, P., Bernard, P.: Key Factors of Success of CRM Software in the Car Industry. International Journal of Research in Engineering and Technology (IJRET) 2(2) (2013)
6. Lambert, M.D.: Customer relationship management as a business process Management. Journal of Business & Industrial Marketing 25, 4–17 (2010)
7. Manoj, P., Pankaj, S., Kirti, K., Santosh, K., Abhishek, G., Devesh, A.: CRM in Indian banking Sector: exploring the critical success factors. International Journal of Business and Emerging Markets 1, 282–295 (2009)
8. Karakostas, B., Kardaras, D.: The state of CRM adoption by the financial services. Journal of Information & Management 42, 853–863 (2005)
9. Ivens, B., Mayrhofer, U.: Les facteurs de réussite du marketing relationnel, revue décision marketing, juillet (2003)
10. Greenberg, P.: The impact of CRM 2.0 on customer insight. Journal of Business & Industrial Marketing 25, 410–419 (2010)
11. Elmuti, D., Grunewald, J., Abebe, D.: Consequences of Outsourcing Strategies on Employee Quality of Work Life, Attitudes, and Performance. Journal of Business Strategies 27, 178–198 (2010)
12. Love, P., Edwards, D.-J., Standing, C., Irani, Z.: Beyond the Red Queen syndrome: CRM technology and building material suppliers. Engineering Construction and Architectural Management 16, 459–474 (2009)
13. Zaynab, M.: CRM implementation in retail banking in Nigeria - A structured literature review. Master's Thesis, Cranfield University, United Kingdom (2007)
14. Levitt, T.: After the sale is over. Harvard Business Review, 87–93 (September-October 1983)

A Journey from Big Data to Smart Data

Fernando Iafrate

Abstract. Nowadays Decision is a matter of information (related to the data), knowledge (related to people & processes) and timing (capacity to decide, act and react in the right timing). The huge increase of the data volume, velocity but also format (unstructured data like: blogs, logs, video…) generated by the "digitalization" of our world modifies radically our relationship to the space (in motion) and time (all the time) dimension and by capillarity, the enterprise vision of the performances monitoring and optimization. This paper introduces via three use cases, how the closed loop between Big Data & Smart Data embedded into the business processes can help the enterprises facing this new challenge.

Keywords: Big Data, Smart Data, Smart Data, Operational Performance Management (OPM), Operational Control Center (OCC), Yield Management, Bid Price, Business Process, Zero Latency Organization, Key Performance Indicator (KPI), Business Intelligence (BI).

1 Introduction

The concept of Big Data usually covers four notions (called the 4V): the data Volume, the data Velocity, the data Veracity, and the data Value, we decided to enhance this approach by splitting the 4V in two subset: Data Volume & Velocity for the Big Data (where we manage the real time information as the "actual" state), and data Veracity & Value for the Smart Data (where we manage the cross functional analytical information as the "future" state), the merge of these two concept called "closed loop" in this paper is the corner stone of our three use cases. Most of the enterprises are struggling with the management of their data, every day; the Digital World is generating new sets of data coming from different sources (the web is the main one) in different formats structured (can be directly

Fernando Iafrate
Senior Manager of the Business Intelligence & Data
Architecture Domain Disneyland
Paris (France)

P.-J. Benghozi et al. (eds.), *Digital Enterprise Design & Management*,   25
Advances in Intelligent Systems and Computing 261,
DOI: 10.1007/978-3-319-04313-5_3, © Springer International Publishing Switzerland 2014

loaded in a relational database) and unstructured (cannot be directly loaded in a relational database, like images, transaction logs, blogs…), lot of these data are holding "noise" (information or meta data having low or no real value for the enterprise).The purpose of our Smart Data (thanks to the analytical skill & tools) is to hold the valuable data of the enterprise and use them to interact in "real time" (most of the time in the transactional process activity and timing) using Business Intelligence applications.

2   A Journey from Big Data to Smart Data

As a closed loop enabled by automated Business Intelligence apps and organization in order to manage & monitor the performances of the enterprise, enabling a Zero Latency Organization.

2.1   What Is?

Big Data (according to the Gartner): is high-volume, high-velocity and high-variety information assets that demand cost-effective, innovative forms of information processing for enhanced insight and decision making.

Smart Data: is in fact a subset of data (Big or not) valuable for the enterprise and cross functional.

   Zero Latency Organization: is an organization where the decision to action latency is directly linked to the process latency you are monitoring/optimizing, it will imply the right "decision" KPI's (key performance indicators related to the business process), Business Intelligence tools (including the capacity to manage the right data velocity) and an organization in a decision for action processes.

Fig. 1 from Big Data to Smart Data

2.2   What For

The Digital World, is in reality more or less about managing "Data" (all kind of data), if you really want to have a serious advantage in this arena, you will need to have a strong skill on Data Management but also the right organization, processes and associated Business Intelligence apps in order to leverage the full value be- hind those data.

   The main goal is to move from a data organization (struggling with the data management) to a learning organization (leveraging all the value behind the data, with the right processes and organization)

2.3   Key Issues

What are the factors that lead to successful strategic deployment of information management in a context where we generate every day more data than what we can really manage/analyze without a strong Business Intelligence organization, au- tomated processes & Business Intelligence apps?

2.4   What You Need to Know

We built a world-class Business Intelligence & Data Architecture environment, which provides customer-focused (before, during and after the visit) strategic and "real-time" (align with the business processes)   insight to a broad set of users, but also automated interaction driven by Business Intelligence apps connected to the "Big and Smart Data" analytical processes & tools. Business activity is predicted and continuously monitored against key performance indicators (KPI's).

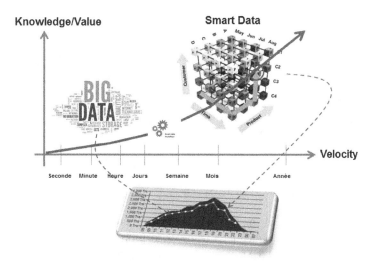

Fig. 2 from Big Data to Smart Data a "closed loop"

We have created a "closed loop" between the "Big Data" (for the real time information showing the current state) and the "Smart Data" (analytical model showing the expected state) in order to monitor our activities. By linking the two data concept (Big Data & Smart Data) we moved form a probabilistic forecast (issue from the Smart Data) having a certain level of probability to a more deterministic approach where the "current state" (extracted from the Big Data) is compared in "real time" (timing linked to the target process) to the "expected state" (forecast generated by the Smart Data) in order to monitor and optimize the current activity and the impacts of the decisions/actions. At the end of the day, the Big Data are then filters and sorted… (The map reduce model), included in the Smart Data (models) in order to enrich them, this closed loop generates an auto learning process, which is the basement of our marketing and sales knowledge management solution.

3   Use Cases

As the context has been defined, I would now like to introduce three cases of application in the enterprise of Business Intelligence applications linked to BIG & Smart Data.

–   The customer relational process (discovery phase)
–   The selling processes (price & availability phase)
–   The operational processes (customer experience in the parks)

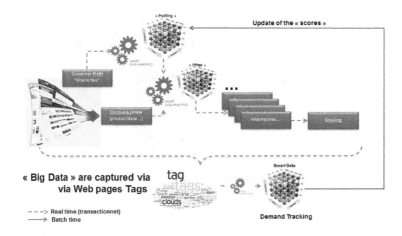

Fig. 3 Use case #1: Demand Tracking System

In this case, we have implemented an "auto learning" solution enabled by a Business Intelligence applications in order to propose on our E-commerce web site, the right product to the right customer at the right time.

This solution is dealing with:

- Real time information coming from the web site (feeding the "on the fly" scoring process)
- Predefine scores and products list modelized by customer segment enabling the "eligibility" (what product fits the customer request) product list for the scored segment.
- The solution generate thousands of price and availability check in order to propose best offers & alternatives to the customer based on the current Bid Price (the bid price, is the minimum expected revenue in the request context: Arrival date/ Length of the stay/ Patry Mix / Hotel Inventory, ~2,8 Millions of bid price point are indexed in real time by using the remaining hotel capacity, more an inventory is constrainted, higher is the bid price)
- The solutions are in an auto-learning mode enabled by the demand tracking "Smart Data" wich recalculate every day the score variance... based on the daily data captured from the web site (all the traffic is modelized, even if it generates no bookings)

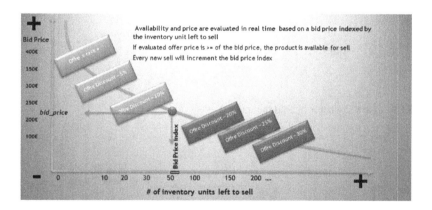

Fig. 4 Use case #2: Bid Price concept for Dynamic Yield Management

The bid price concept aim to validate the product availability based on minimum expected revenue (the bid price) directly indexed with the inventory (room in this case) left to sell.

The expected product value (link to the customer request for an Arrival date/ Length of the stay/ Party mix/ Hotel inventory, is compared to the bid price (minimum expected revenue for this context), if the product value is >= to the bid price, the product will be available for sales, else the product will be consider as unavailable (even if we still have free inventories).

Every inventory update (hotel room in this case) by a sale (reduce availability) or a cancel (increase availability) will impact the bid price index, and by capillarity will reference a new bid price dynamically, this new bid price will then be compare with the requested product value, in order to enable of disable the product availability. This mechanism will secure the intraday optimization by reacting in "real time" to the customer behavior (not only to what was forecasted the previous night in term of customer demand)

The bid price is generated by our yield management solution every day, we generate more than 2.8 Millions of bid price points (number of inventories x forecast horizon), see next figure (5) for more details.

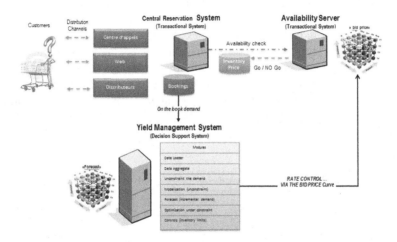

Fig. 5 Use case #2: what is behind the magic?

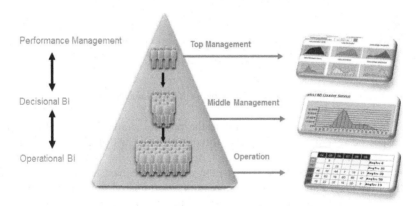

Fig. 6 Use case #3: the "Data" align connect all the level of the enterprise

In order to leverage the bid price table produced (every night) by the Yield Management solution (as an output of the demand forecast by product value), we had to implement an Availability Server. The Availability Server is embedded in the selling transactional process and activated during the availability and price check, the architecture is scalable and can hold thousands of availability & price request per minute, in order to support the organic growth of the digital activity on our E-Commerce web site.

The OPM (Operational Performance Management) aim to align and connect all the level of the enterprise by managing in "real time" (align with the timing of each one of the target processes) the operational performances of the enterprise.

The onsite operational performance of the enterprise will have a direct effect on the customer experience, and our company is the place where the "dreams comes true". In order to monitor and manage this operational performance, we have implemented an OPM (Operational Performance Management Solution).

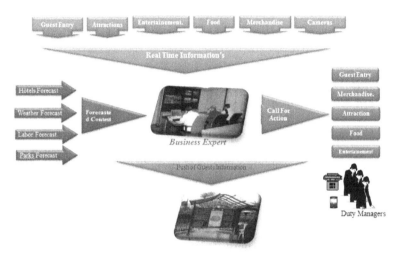

Fig. 7 Use case #3: the OCC (Operational Control Center) [Grundstein & Rosenthal-Sabroux, 02]

The OPM (Operational Performance Management) is a set of Business Applications publishing "real time" (link to the target process timing) information and alerts as KPI's for the operational processes, but also the right organization to leverage that information.

The OCC (Operational Control Center) is an example of an organization leveraging (via business experts of each one of the operational domains) the Big to Smart data information for the monitoring and the optimization of the operational performances of the enterprise.

4   Conclusion

Three types of latency (Based on HACKATHORN, 2004)

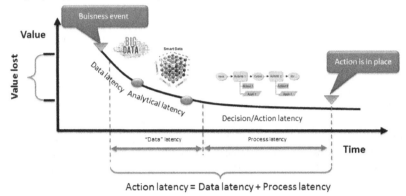

Fig. 8 Data value is a matter of timing and alignment with the business processes [Hacka-thorn, 01]

4.1   What Are the Main Features Links to the Big Data to Smart Data Closed Loop

- **Automate analysis.** In order to automate day-to-day operational decision-making, organizations need to be able to do more than simply present data on a dashboard or in a report. The challenge is turning real-time data into something actionable. In short, businesses need to be able to automatically interpret data, dynamically, in real time. What this means in practice is the ability to compare each individual event with what would normally be expected based on past or predicted future performance. Smart Data must understand what normal looks like at both individual and aggregate levels and be able to compare individual events to this automatically.

- **Forward looking.** Understanding the impact of any given event on an organization needs to be forward looking (predictive BI)

- **Process oriented.** To be embedded within a process in order to make the process inherently smarter requires Smart Data need to be process-oriented. This doesn't mean that the process has been modeled with a business process management tool. Actions can be optimized based on the outcome of a particular process, but the process itself may or may not be explicitly defined.

- **Scalable solutions.** Scalability is naturally a cornerstone because it is based on event-driven architectures. This is critical because event streams can be unpredictable and occur in very high volumes.

- **An intelligent organization.** To give to the enterprise a tactical advantage in the business arena. Concern with transforming your data into an accessible store of high-value information and distributing the right information at the right time to the right person for the right decision. It's easy to see the correlation between the quality of the Business Intelligence and the success of the operation (those who comprehend and act quickly upon relevant facts have advantages over those who do not)

4.2   What Are the Main Risks of Not Implementing a Strong Business Intelligence Practice for the Enterprise?

As the world becomes more digital every day, which, by extension, creating a new stream of data types and associated speed and volume, the company that currently struggling to manage their data will be in a worse situation in the near future

References

Warden, P.: Big Data Glossary, Editions O'Relly

Smolan, R., Erwitt, J.: The Human Face Of Big Data

Business Intelligence for the Enterprise, Mike Biere, Edition IBM presse

Big Data: A Revolution That Will Transform How We Live, Work, and Think; Viktor Mayer-Schonberger and Kenneth Niel Cukier, Hmhbooks (2013)

Learning From Data by Yaser S. Abu-Mostafa, Malik Magdon-Ismail and Hsuan-Tien Lin, AMI Books (2012)

Yield et Revenue Management, Jean-Paul Sinsou, Edition ITA presse

Cross, R.G.: Yield Management. Broadway Books (1997)

Talluri, A.T., Ryzin, G.J.: The Theory and Practice of Revenue Management. International Series in Operations Research & Management Science. Springer (2005)

Hackathorn, R.: The BI Watch: real-Time to Real-Value. DM Review 14(1) (January 2004)

Grundstein, M., Rosenthal-Sabroux, C., Pachulski, A.: Reinforcing Decision Aid by Capitalizing on Company's Knowledge: Future Prospects. In: EJOR 2002 (2002) (to appear)

Harnessing Multimodal Architectural Dependencies for IT-Business Alignment and Portfolio Optimization: A Statistical Approach

Zenon Mathews, Livio Kaeslin, and Bernhard Rytz

Abstract. With the growing importance of architectural considerations over the last decade, larger Enterprise Architecture DataBases (EADBs) have enjoyed accumulation of enterprise data over years. Enterprise data documenting IT-applications, interfaces, data ownerships, business function implementation and usage etc. have been modelled based on various underlying metamodels and using different EA-tools. However, harnessing this data for business benefit is often hard and widely performed manually using expert knowledge. We here propose a statistical method for harnessing the EADB knowledge to infer indirect dependencies between IT-applications and projects. Our approach provides key insights into otherwise unseen multimodal dependencies in the IT landscape and provides a quantifiable methodology for optimizing the IT development plan and portfolio management. We validate our method in the highly integrated IT landscape of the Swiss Federal Railways, one of the world-leading rail transportation providers, and show how business decision making can directly be supported using our approach.

1 Introduction

Over the last decade, Enterprise Architecture (EA) has provided transparence to IT landscapes and made the IT landscape of high-performance enterprises more manageable [1]. Nevertheless, much of the EA data accumulated over the years, documenting different architectural layers including business function, data, application and technology, often remain unused when making architectural decisions with business impacts [1, 2]. The highly complex application architecture with large

Zenon Mathews · Bernhard Rytz
Enterprise Architecture, SBB Informatik AG, Swiss Federal Railways
e-mail: {zenon.mathews,bernhard.rytz}@sbb.ch

Livio Kaeslin
SBB Passenger Traffic
e-mail: livio.kaeslin@sbb.ch

P.-J. Benghozi et al. (eds.), *Digital Enterprise Design & Management*,
Advances in Intelligent Systems and Computing 261,
DOI: 10.1007/978-3-319-04313-5_4, © Springer International Publishing Switzerland 2014

number of interfaces and a rich project portfolio in bigger enterprises prohibit the manual in-depth exploitation of the EA databases for the above use [10]. As a result, many architectural decisions and project portfolio optimizations are based on qualitative assessments and locally optimal expert suggestions. This approach, although very valuable, can subsequently lead to the development of IT solutions to possess undiscovered dependencies to other developments within the organization. In the long-term, this leads to functional and data redundancies, making the enterprise less and less agile with time.

Indirect repercussions of cancellation or postponement of a specific IT-project on other ongoing projects usually remain unknown. We provide a quantifiable measure to assess such indirect impacts and help optimizing the project portfolio from the enterprise perspective. We benchmark our approach in the Swiss Federal Railways (SBB, for the German Schweizerische Bundesbahnen) IT landscape and compare our results with expert assessments. Our results demonstrate the usefulness of the approach in a high-performance transport industry, that provides one of the densest and most punctual public transportation services world-wide.

IT-application portfolios of larger enterprises often have hundreds of applications with thousands of interfaces [10]. Conventional approaches of managing the high complexity in such application portfolios include the division of the landscape into separate functional domains [3, 4]. For example, the SBB application landscape consists of more than 1000 applications divided into several functional domains (figs.1, 2). Each domain covers a subset of applications similar in their (business) function. Each domain is managed by a domain architect, who drives the development plan of the domain and coordinates the different project efforts of the domain, considering both IT architectural principles and business perspectives. Domain expert knowledge is used to optimize the development plan over the coming years. For individual domains that are architectonically complex, domain knowledge itself is not always sufficient for a complete dependency analysis. Moreover, as individual domains are optimized mostly locally, inter-domain dependencies are only addressed in very obvious cases. This approach leads to dependency issues with the growing pace of beyond-domain integration of today's IT-landscapes.

A development plan considering the global view of the enterprise should therefore consider dependencies that go beyond domain boundaries. Furthermore, the so called what-if analysis is considered very helpful in early project stages for recognizing dependencies to other projects both inside and outside the given domain [1]. Portfolio optimization and financially sustainable development plans therefore need to apply methods that enable an enterprise-wide dependency analysis of projects. This is however a challenging task as dependencies in larger IT-application portfolios exist at different levels; e.g. the TOGAF reference metamodel [6] itself has relationships between the business, application, data and technology architectures. With increasing importance of enterprise architecture modelling for sustainable growth of the enterprise-wide IT-systems, the maturity of available information about such dependencies has grown remarkably in the recent years [1]. Nevertheless, effective mathematical techniques are rarely used to harness such multimodal dependency information.

We here propose a statistical sampling technique, the Markov-Chain-Monte-Carlo Gibbs sampling [11], to infer dependencies in the IT-landscape and allow project portfolio planners to recognize and align dependent projects. Our approach can harness multimodal dependencies (i.e. dependencies at multiple architectural levels) in the information system landscape and can also cope with the common cyclic dependency issue. We apply and demonstrate the usefulness of the method in the SBB IT-landscape. Our proposal can directly help cut project costs and increase business effectiveness.

1.1 Related Work

EA allows the management of information in enterprises with large information systems. EA is often model-based and diagrams represent systems and their inter-dependencies. In the recent years, formal analysis of such EA models have been shown to be very powerful in many subdomains such as security, portfolio optimization and ensuring IT agility [2]. Strong proposals for statistical approaches to dependency inference in EA models have already been proposed [5, 9, 8]. Such approaches mostly consider unimodal dependency scenarios in information systems. Moreover, the assumption of acyclic dependencies often cannot be fulfilled in many large integrated information systems as inference in Bayesian networks does not allow cyclic dependencies [5].

Our approach, on the contrary, is more powerful in scenarios with dependencies at multiple levels of the information system (application, business function, data and technology). Moreover, the Gibbs sampling technique for Markov Chain Monte Carlo allows for the existence of cyclic dependencies in the EA model [11].

1.2 The SBB Information Systems Landscape

The SBB is the national railway company of Switzerland, which is a special stock corporation with shares exclusively owned by the Swiss Confederation. The SBB has five major divisions: passenger traffic, freight, real-estate, infrastructure and core services. With almost 1 million passengers per day, one of the densest national railway networks, and almost 90% of arrivals within 3 minutes of scheduled arrival time, the performance index of the SBB is known to be world-leading.

The SBB IT landscape consists of above 1000 applications, thousands of interfaces between applications and about 170 IT projects per year that constantly develop the application landscape (see fig. 1, 2). Owing to this dynamic development scenario and to enable business agility, the team of architecture and quality ensures the design and update of development plans. This process is performed several times per year for each of the 13 IT domains in consultation with the IT finance department.

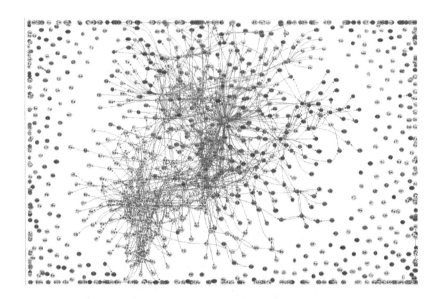

Fig. 1 SBB IT Applications (color coded nodes for IT domains) and corresponding interfaces (directed edges)

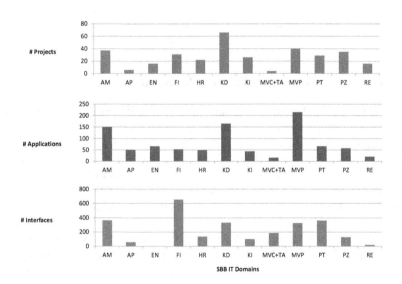

Fig. 2 SBB IT domains at a glance

2 Methods

2.1 From Metamodel to Dependency Graphs

SBB employs the metamodel shown in figure 3. We use the following entities and relationships from the metamodel (translated from German):

- "Anwendung": Application: a high level logical representation of a collection of software applications, usually implementing a specific business functionality. We consider interfaces between applications as dependencies (interfaces are modelled as messages "Nachricht")
- "Projekt": IT project: an approved IT project from the IT portfolio that has a direct impact on one or more applications, which we consider as dependencies.
- "Geschäftsfunktion": business function: an application either implements a business function or uses it. Implementation and usage of business functions are modeled as dependencies in our analysis.
- "Plattform": platform: an application is run on a platform. We consider an application as dependent on the platform it runs.
- "Geschäftsdatentyp": business data type: an application can own or use one or several business data type. Also this data ownership or usage are used to model dependencies in our analysis.

We therefore consider multimodal dependencies at the data level, business functionality level and at the platform level. By considering these multimodal dependencies in the enterprise database, we form the so called *dependency graph*, as shown in figure 4. Such dependency graphs are used by the Markov Chain Monte Carlo algorithm to deduce indirect dependencies between applications and IT-projects. Such indirect dependencies are useful for coordinating the development plan and making use of synergies. Indirect dependencies are often hidden in early project phases and cause much effort and cost in later phases as the projects have already advanced further.

2.2 The Markov Chain Monte Carlo (MCMC) Gibbs Sampler

Let S be a finite set of states $S = s_1, s_2, \ldots, s_n$. A finite, discrete time, first order Markov Chain is a stochastic process, which is a sequence of random variables $X = ((X_t)_{t=0,1,\ldots}$ with values in S and with the property:

$$P(X_{t+1} = s_{j_{t+1}} \mid X_t = s_{j_t}, X_{t-1} = s_{j_{t-1}}, \ldots, X_0 = s_{j_0})$$
$$= P(X_{t+1} = s_{j_{t+1}} \mid X_t = s_{j_t}) \tag{1}$$

The *transition probability* is the probability that the random variable with time-index t changes from state s_i to state s_j. This property means that the probability to reach a certain state in the next step only depends on the actual state and is unaffected by the anterior states. This probability is p_{ij} and is defined as

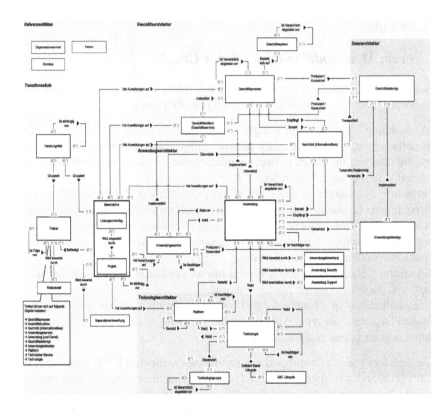

Fig. 3 SBB IT Metamodel: We use the relationships between applications (Anwendung), Projects (Projekt), platforms (Plattform), business data type (Geschäftsdatentyp) and business functions (Geschäftsfunktion) in the MCMC algorithm.

$$p_{ij}(t) := P(X_t = s_j \mid X_{t-1} = s_i) \tag{2}$$

A Markov chain is *homogeneous* if all transition probabilities are independent of time. Often a *transition matrix* is used to indicate the transitions probabilities from every state to every other.

The probabilities of the initial states are given by the initial state probability vector. The stationary distribution is a probability vector that describes the state probability vector of the sought solution of the problem. Convergence to the stationary distribution is achieved if the following conditions are fulfilled:

• *irreducibility*: a Markov Chain is irreducible when each state can be reached from any other state in a finite number of steps
• *aperiodicity*: a Markov Chain is aperiodic if no state is reached only periodically
Convergence of a homogeneous, irreducible and aperiodic Markov Chain is mathematically guaranteed and there is just one stationary distribution to which the chain converges.

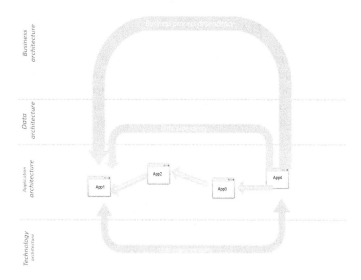

Fig. 4 Schematic of the used multimodal dependency graph constructed from the SBB EA-model

In our case, we consider the state of the dependency graph of the IT-landscape in each Markov chain step. The state of the dependency graph is simply the joint states of each application in the landscape. Each application can thereby have one of the two following states:

- A, meaning that the application undergoes a change
- B, meaning that the application does not undergo a change

Given the dependency graph G, we define the probability that an application will have the state A in the next step as follows:

$$P(State = A \mid G) := \frac{\#altered\ in - ports}{\#in - ports} * c \tag{3}$$

where, a constant $c(0 < c < 1)$ ensures the irreducibility of the chain. An *altered in − port* thereby means an interface/port leading from an application that is in state A to the current application. Further, it can also be shown that the above defined Markov Chain is aperiodic. We use the Gibbs Sampling, a so called Monte Carlo method [11], to find the stationary distribution of the above Markov Chain.

Let A_k^i be the status of the $k - th$ application in the $i - th$ iteration step. Given the initial state of all m applications and their states in the $i - th$ iteration step:

$$A^i = (A_1^i, A_2^i, \ldots A_m^i) \tag{4}$$

the Gibbs method uses the state probability as in equation 3, and samples the next state of each application for the next step as

$$A^{i+1} \sim P(A_j \mid A_1^i, A_2^i, \ldots, A_{j-1}^i, A_{j+1}^i, \ldots, A_m^i) \qquad (5)$$

until the stop criterion is reached. The stop criterion is defined by ε, which is a small enough euclidean value, that indicates the almost steady average state of the dependency graph. It is important to note that the computed probabilities here represent the probability of dependence between entities (applications and projects), and not for example how dependent an entity is dependent of another one.

2.3   Tools

The ©MEGA Suite is a widely recognized multiuser tool for enterprise architecture modelling, governance, risk and compliance tasks. The SBB-IT employs the ©MEGA tool for EA modelling. All domain architects model the attributes of the applications, projects, interfaces (ports) between applications of their corresponding domains respecting the SBB-wide metamodel and the ©MEGA tool. We here use the ©MEGA repository for the dependency analysis. We use the prefuse visualization toolkit [7] to visualize the dependency graph of the IT-landscape and the indirect dependencies between several IT projects.

3   Results

3.1   Project Indirect Impacts in the Application Landscape

We consider a single IT undertaking of the SBB IT, removing a legacy applications from the SBB IT landscape. The direct impacts are currently not yet documented in the EA-DB but are widely known by the domain experts. Nevertheless, the indirect impacts of the project are crucial yet unknown. We use the proposed MCMC algorithm to compute the indirectly impacted applications of this undertaking. The results of this "what-if" analysis are shown in figure 5. (the considered project is 'SYFA-Ablösung'). This result is invaluable for the project and the domain architect, and can be used as a guide when coordinating the development plan of the domain. Analysis with such pseudo-projects (i.e. projects that in reality do not exist in the portfolio) can be easily performed using our tool, to analyze potential impacts in the application landscape of a change to a particular application.

Currently, following an strategic IT decision, the operation of the Mainframe host platform is being relocated from a local provider to foreign provider. A major concern during this relocation project was that iterative function calls from non-host applications to host applications could cause prohibitively high round trip times, which subsequently can be unacceptable to critical applications. As there are numerous applications running on the host platform, with large number of data interfaces to non-host applications, a manual inspection of the potentially impacted

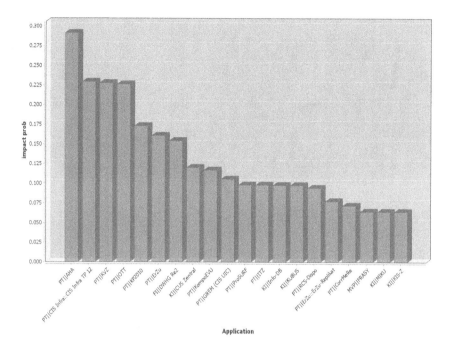

Fig. 5 What-if analysis: indirectly impacted IT-applications of a potential IT project

applications was not possible (thousands of potential candidates). We applied our tool to aid solving this problem. The direct project impacts were applied to the host platform, that was to be relocated. After this we analysed the indirect impacts in the application landscape. The 30 most probable indirectly impacted applications are shown in fig. 6. Our results could help platform architects and application managers to look into specific interfaces and perform tests to check the round trip times. It turns out that interfaces were found using our tool, which were previously not even considered and potentially could be critical after the relocation of the host.

3.2 Inter-Project Dependencies

Going beyond indirectly impacted applications, we are also interested in dependencies between projects. To infer such dependencies, we run the MCMC algorithm pairwise for each project-project pair. This allows to infer the highly probable dependencies as discussed in section 2.2. As a result of such pairwise comparison of projects and subsequent thresholding of dependency probabilities, we can compute the probabilistically highest dependencies between projects. Note that such highly probable dependencies can occur beyond domain boundaries, making the conventional domain oriented development plans insufficient.

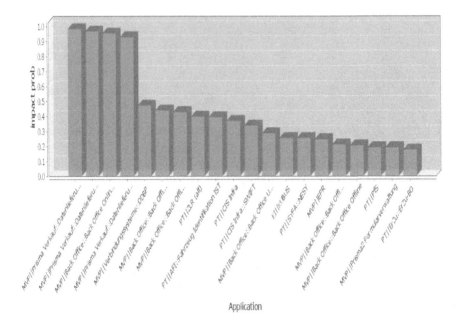

Fig. 6 Indirectly impacted applications of the host relocation project

Figure 7 illustrates the probable dependencies of the pseudo-project, which would remove a legacy application from the SBB-IT landscape. Such what-if analysis are highly valid not only for planning the removal of highly critical and integrated applications but also when introducing new projects into a rich portfolio. In figure 7, the pseudo-project is the filled circle in the center, the concentric circles indicate probability bounds and the colored filled circles indicate other real IT projects. The closer the projects to the pseudo project, the more probable their dependencies. Note that projects from a wide range of domains might be affected by the removal of the above mentioned legacy system. The shaded triangles indicate the common application to all projects inside the triangle, with indirect project impacts on that application.

4 Discussion

Our results serve as an entry point for further investigation of project dependencies and indirect impacts of projects in the IT-landscape. In our experience, the added value has been received very well by the domain experts. We have also compared the results of our approach with expert knowledge to evaluate the quality of the dependency analysis. Our approach could thereby not only confirm known dependencies between projects, but also discover unseen dependencies and indirect impacts.

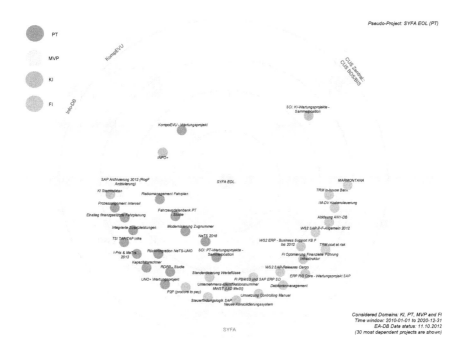

Fig. 7 Using a pseudo-project to estimate the impacts of an application EOL on ongoing IT-projects

Our approach is currently being tested at SBB-IT as an analysis tool to aid domain experts and portfolio planners with the development plan. Our proposal overcomes the restrictions of earlier approaches based on Bayesian inference, where only unimodal dependencies in acyclic dependency graphs could be used. Inclusion of the approach in formal development planning procedure is currently being considered. Furthermore, we are currently investigating the application of the approach in the software change management process, where changes to the current IT-landscape can have indirect impacts. Today, the dependencies between changes are assessed merely using expert knowledge.

Acknowledgments. The authors would like to thank Hans-Jakob Gfeller, Peter Moosmann, Jakob Erber, Stefan Meichtry, Yannis Baillet, Christoph Trutmann, Ugur Asuroglu, Peter Hug, Rudolf Gysi, Oliver Hofer and Hans-Peter Weiss (all at SBB Informatik and SBB Passenger Traffic) for valuable suggestions and discussions.

References

1. Ahlemann, F., Stettiner, E., Messerschmidt, M., Legner, C.: Strategic Enterprise Architecture Management: Challenges, Best Practices, and Future Developments. In: Management for Professionals. Springer (2012)
2. Beyer, M., Lovelock, J.D., Sommer, D., Adrian, M.: Big data drives rapid changes in infrastructure and 232 billion in it spending through 2016. Gartner Research (2012)
3. Evans, E.: Domain-Driven Design: Tackling Complexity in the Heart of Software. Addison-Wesley (2004)
4. Fowler, M.: Patterns of Enterprise Application Architecture. Addison-Wesley Longman Publishing Co., Inc., Boston (2002)
5. Franke, U., Flores, W.R., Johnson, P.: Enterprise architecture dependency analysis using fault trees and bayesian networks. In: Proceedings of the 2009 Spring Simulation Multiconference, SpringSim 2009, pp. 55:1–55:8. Society for Computer Simulation International, San Diego (2009)
6. Haren, V.: Togaf Version 9.1. Van Haren Publishing Series. Bernan Assoc. (2011)
7. Heer, J., Card, S.K., Landay, J.A.: Prefuse: a toolkit for interactive information visualization. In: Proceedings of the SIGCHI Conference on Human Factors in Computing Systems, CHI 2005, pp. 421–430. ACM, New York (2005)
8. Johnson, P., Lagerström, R., Närman, P., Simonsson, M.: Enterprise architecture analysis with extended influence diagrams. Information Systems Frontiers 9(2-3), 163–180 (2007)
9. Lagerstrom, R., Johnson, P.: Using architectural models to predict the maintainability of enterprise systems. In: Proceedings of the 2008 12th European Conference on Software Maintenance and Reengineering, CSMR 2008, pp. 248–252. IEEE Computer Society, Washington, DC (2008)
10. Sessions, R.: Simple Architectures for Complex Enterprises. Best Practices. Microsoft Press (2008)
11. Walsh, B.: Markov Chain Monte Carlo and Gibbs Sampling. Lecture Notes for EEB 581 (2004)

Towards a Model-Driven-Architecture Process for Smart Grid Projects

Christian Dänekas, Christian Neureiter, Sebastian Rohjans,
Mathias Uslar, and Dominik Engel

Abstract. The complexity in electrical power systems is continuously increasing due to its advancing distribution. This affects the topology of the grid infrastructure as well as the IT-infrastructure, leading to various heterogeneous systems, data models, protocols, and interfaces. This in turn raises the need for appropriate processes and tools that facilitate the management of the systems architecture on different levels and from different stakeholders' view points. In order to achieve this goal, a common understanding of architecture elements and means of classification shall be gained. The Smart Grid Architecture Model (SGAM) proposed in context of the European standardization mandate M/490 provides a promising basis for domain-specific architecture models. The idea of following a Model-Driven-Architecture (MDA)-approach to create such models, including requirements specification based on Smart Grid use cases, is detailed in this contribution. The SGAM-Toolbox is introduced as tool-support for the approach and its utility is demonstrated by two real-world case studies.

1 Introduction

One of the key challenges resulting from the Smart Grid vision is to handle complexity in distributed systems [3]. The first step to address this challenge is to structure the overall domain. In this context, the results of the European Standardization Mandate M/490 currently gain momentum, especially the Smart Grid Architecture Model (SGAM) [12]. The SGAM has been developed by members from CEN,

Christian Dänekas · Sebastian Rohjans · Mathias Uslar
OFFIS – Institute for Information Technology, Escherweg 2, 26121 Oldenburg, Germany
e-mail: {daenekas,rohjans,uslar}@offis.de

Christian Neureiter · Dominik Engel
Josef Ressel Center for User-Centric Smart Grid Privacy, Security and Control
Salzburg University of Applied Sciences, Urstein Süd 1, A–5412 Puch/Salzburg, Austria
e-mail: {christian.neureiter,dominik.engel}@en-trust.at

P.-J. Benghozi et al. (eds.), *Digital Enterprise Design & Management,*
Advances in Intelligent Systems and Computing 261,
DOI: 10.1007/978-3-319-04313-5_5, © Springer International Publishing Switzerland 2014

CENELEC and ETSI and considers established domain models (e.g., from NIST and IEC) as well as domain-independent architecture frameworks such as TOGAF. Furthermore, in terms of interoperability dimensions the GridWise Interoperability Context Setting Framework was adopted. As shown in Figure 1, it provides the means to express domain-specific viewpoints on architecture models by the concepts of *Domains*, *Zones* and *Interoperability Layers*, which shall be briefly introduced in the following.

The *Domains* regard the energy conversion chain and include: *generation* (both conventional and renewable bulk generation capacities), *transmission* (infrastructure and organization for the transport of electricity across long distances), *distribution* (infrastructure and organization for the distribution of electricity to the customers), *DERs* (distributed energy resources connected to the distribution grid) and *customer premises* (both end users and producers of electricity, including industrial, commercial, and home facilities as well as generation in form of, e.g., PV generation, electric vehicles storage, batteries, and micro turbines).

The hierarchy of power system management is reflected within the SGAM by the following *Zones*: *process* (physical, chemical or spatial transformations of energy and the physical equipment directly involved), *field* (equipment to protect, control and monitor the process of the power system), *station* (areal aggregation level for field level), *operation* (power system control operation in the respective domain), *enterprise* (commercial and organizational processes, services and infrastructures for enterprises), and *market* (market operations possible along the energy conversion chain).

Finally, as it constitutes a major requirements towards distributed systems the SGAM defines *Interoperability Layers*. These cover entities ranging from business objectives to physical components to express the respective architecture viewpoint. Like proposed by TOGAF, interrelations between concepts from different layers shall ensure traceability between architecture properties.

Furthermore, the Mandate M/490 proposes a methodology for use case management also including a template for use case descriptions [13]. It is suggested by the original reports to combine this methodology with the SGAM in order to create architecture models addressing the requirements elicited in context of appropriate use cases. However, the original purpose of the SGAM is to identify standardization gaps. Therefore, it has to be adopted to be suitable for a genuine model-driven architecture process. In [1] and [14] gaps for applying the SGAM beyond its original purpose have been identified. One issue that has been pointed out by both is its missing formalization. This gap will be addressed in this contribution by proposing Unified Modeling Language (UML) meta-models in order to formally describe SGAM-based architectures.

The remainder of this contribution is structured as follows: The overall approach is presented in Section 2. Two case studies conducted in this context are afterwards outlined in Section 3 and provide further details on possible application contexts and design decisions. The contribution is concluded in Section 4, which also outlines future work.

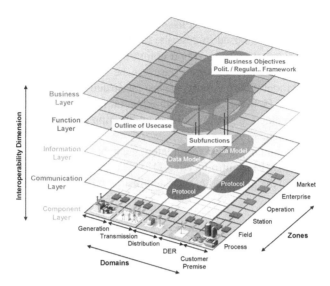

Fig. 1 Smart Grid Architecture Model (SGAM) [12]

2 Approach

As outlined in [3], it is challenging to handle the complexity of distributed systems, especially if they provide critical infrastructures, as in case of Smart Grids. A broadly accepted approach to cope with complexity during the engineering process is the concept of Model Driven Engineering (MDE), which serves as an umbrella term for model-based approaches. Considering the Smart Grid as an interdisciplinary System of Systems (SoS), the architecture-focused concepts of Model Driven Architecture® (MDA®) [9, 8] appear suitable to analyze, decompose and develop Smart Grid related systems. In contrast to Model Driven Software Development (MDSD), MDA focuses on the structuring of specifications rather than on the generation of implementation artifacts.

The utilization of model-based approaches to understand, analyze and design Smart Grid systems has been investigated and deemed valuable by Lopes et al. [7]. The authors strongly rely on the NIST Smart Grid Conceptual Reference Model [6] for the decomposition of Smart Grid systems. This is remarkable, as the NIST Reference Model has been one of the key influencing factors for the development of the SGAM [12].

Similar to [7] the presented work considers model-based approaches to analyze and decompose Smart Grid Systems. Beyond that it aims at the application of these concepts to the engineering task. Hence, initially a fundamental engineering process has been formulated. This process reflects the top-down engineering concepts and consists of a System Analysis Phase, a System Architecture Phase and a Design and Implementation Phase. The deliverables of these phases are the single MDA viewpoints as described in [8].

In more detail, the System Analysis Phase first delivers the Computation Independent Model (CIM), which is used to describe the functionality of a system without mentioning the implementation. Having a common understanding of the intended functionality, the System Architecture Phase can be started to elaborate the Platform Independent Model (PIM). Finally, the last phase delivers the Platform Specific Model (PSM) and the Platform Specific Implementation (PSI). This phase reflects agile ideas and is executed iteratively, with each iteration delivering a vertical slice of the architected system.

Having this development process formulated, it can be aligned with the Use Case Mapping Process (UCMP) as introduced in [12]. The mapping of the six individual tasks of the UCMP to the proposed process is illustrated in Figure 2.

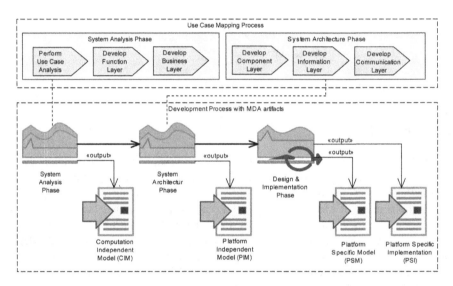

Fig. 2 Mapping of the UCMP to the basic MDA artifacts

3 Case Studies

In the following, two case studies will be outlined, which involve the model-driven architecture development for Smart Grid solutions. This way, the application context for the MDA-approach outlined in the previous section and design descisions regarding conceptual meta-models and tool support will be outlined. Finally, the concepts and experiences from the case study will be compared with each other in order to identify common concepts and approaches as well as differences that motivate future work in this area.

3.1 INTEGRA

In the "Smart Grids Model Region Salzburg"[1] various Smart Grid systems were realized as individual research projects. Currently, the project INTEGRA focuses on the secure and stable operation of the various, mutually influencing systems. In a first step, it is necessary to gain a comprehensive view of the single systems and their interactions. Hence, the MDA-based approach as proposed in Section 2 is utilized for reengineering and modeling selected projects in context of the SGAM.

One of the selected projects, named "Häuser als interaktive Teilnehmer" (HiT)[2], realizes a Smart-Grid-ready energy supply system (thermal and electric energy) for a building complex[3] with 129 flats on 11.000 square meter. A local "Energy Management Center" (EMC) controls the operation of the integrated energy producers and consumers and provides some flexibility to the Distribution System Operator (DSO).

For modeling this system in context of the SGAM, a Domain Specific Language (DSL) was formulated in a first step. This DSL utilizes the SGAM structure and is implemented as a UML-based meta-model. It consists of functional and structural components and their respective horizontal and vertical relations. The basic structure of the meta-model is illustrated in Figure 3. To keep the illustration compact, it does not comprise the attributes of the individual elements.

The depicted meta-model reflects the layered structure from the SGAM and is aligned with the elements from the modified Use Case Template (UCT) that is suggested to be used for the collection of use cases in [13]. The hierarchy between the use cases follows the ideas from [13]. A High Level Use Case (HLUC) invokes a number of Primary Use Cases (PUC). Each of them is composed of different scenarios, which are made up of a number of Use Case Steps that describe the functionality in detail. If a Use Case Step includes an information exchange, an appropriate Information Item element has to be created and associated to the Use Case Step. These Information Item elements can be used in the consecutive steps of the Use Case Mapping Process, e.g., to identify adequate standards or to model Information Object Flows in a consistent manner.

Following the ideas of the MDA, the described functionality is to be mapped to a technical viewpoint, which states the model transformation from the CIM to the PIM. Thus, if a logical actor is realized by a physical component, an adequate mapping has to be done. This takes place during the development of the component layer. Hereby, the logical actors can be mapped either to physical devices ("component") or to applications that are hosted by physical devices. This mapping is represented by a "trace" relation in the meta-model. On the component layer level the relations between individual components are expressed by unspecified ICT or Electric Associations. These relations are to be refined in the Information Layer (Data Model Standard and Information Object Flow) and the Communication Layer. The

---

[1] http://www.smartgridssalzburg.at
[2] Meaning "Buildings as interactive participants"
[3] http://www.rosazukunft.at

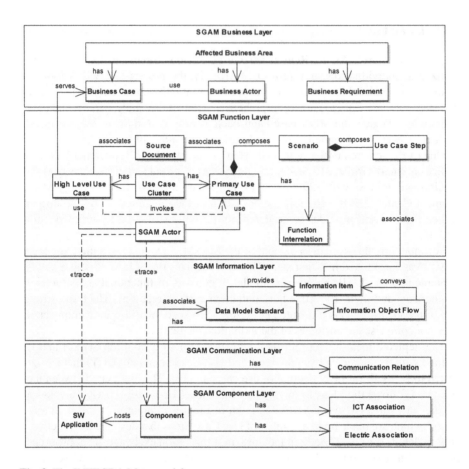

Fig. 3 The INTEGRA Meta-model

introduced Communication Relation comprises adequate attributes (not visible in the illustration) to specify the communication protocol and technology in detail.

Taking the described DSL as a basis, a UML-based toolbox[4] was developed as an extension to the "Enterprise Architect"[5] modeling tool. It was implemented by utilizing the integrated Model Driven Generation (MDG) functionality. In addition to the DSL, the toolbox comprises a number of templates that reflect the UCMP.

During the application of the toolbox some valuable experiences were made. First, it became clear that the analysis of the functionality constitutes a huge part of the total workload. To efficiently deal with this complexity, a more explicit definition of the hierarchy and the relations between the corresponding elements would be useful.

---

[4] The SGAM-toolbox is publicly available for download at
http://www.en-trust.at/SGAM-Toolbox
[5] http://www.sparxsystems.com

Another experience addresses the interrelations between individual HLUCs. For example, the central HLUC "Optimize local energy" is related to the HLUC "Control power of CHP". It would be useful to explicitly define the horizontal interactions between the single HLUCs.

Also, the Business Layer needs some more attention. Due to the technology-driven character of the research projects, the underlying business cases and processes are not completely described and hence, a sufficient evaluation could not take place.

Another aspect is the integration of non-functional requirements. Even if the issue "Security" is discussed in detail in the Information Security Working Group report [11] it could be valuable to integrate the elicitation of non-functional requirements explicitly in the System Analysis Phase.

Besides the experiences mentioned above, the usage of an MDA-based toolbox turned out very helpful during the reengineering process. It was suitable to enhance the Requirements Engineering Process (System Analysis) and to gain a comprehensive view on the system's functionality and structure.

3.2 DISCERN

The EU FP7 project DISCERN[6] aims at providing a common view on Smart Grid solutions based on conceptual specification as well as its application in context of demonstration sites. The distribution grids represent the project's focus area, as this part of the power system faces considerable changes regarding the amount of decentralized generation. Thereby it represents an active part of the power system, resulting in new challenges for the DSOs. In this light, documenting the solutions in a consistent way shall provide the ability to determine the optimal amount of intelligence provided by Information and Communication Technologies (ICT) in the distribution grids, e.g., in terms of cost-effectiveness. Key Performance Indicators (KPI) shall be designed in order to express the solutions properties and provide the basis for comparative assessments between Smart Grid solutions.

To compare these in a holistic way, two dimensions are considered in DISCERN. The initial requirements elicitation of potential solutions is supported by a use case template based on the approach proposed by the Sustainable Processes Working Group in context of the M/490 Mandate [13]. Their architectural properties are afterwards documented by the means of SGAM models [12]. The DISCERN approach thereby builds on the proposal of the Smart Grid Coordination Group (SGCG) to align the creation of use cases and architecture models.

Beyond the initial scope of standardization gap analysis, the DISCERN project aims at applying use case descriptions and architecture models as the conceptual basis for the creation of KPI. To achieve this goal, the entities of both use case and SGAM model were formalized. The resulting model is shown in Figure 4.

To support the application of use cases and architecture models in DISCERN, several adjustments had to be done to the SGCGs results. The central concept in

---

[6] http://www.discern.eu

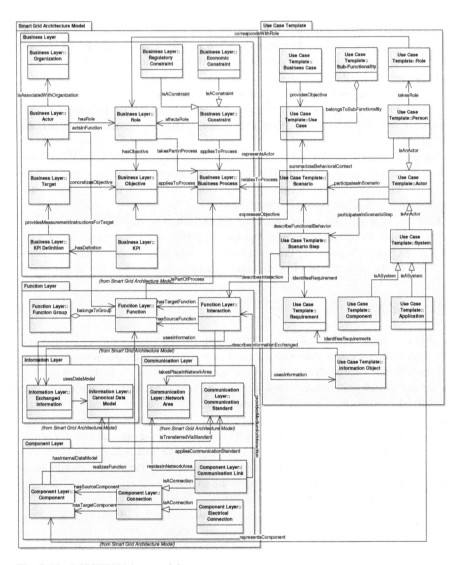

Fig. 4 The DISCERN Meta-model

DISCERN are sub-functionalities, which are used to structure objectives and approaches of a DSO regarding Smart Grid solutions in distribution grids. Use cases are organized under sub-functionalities and express business actors, objectives and business cases as well as the solutions behavior and technical requirements (e.g., technical actors and information objects) in context of scenarios.

For documentation of use cases in DISCERN, a use case template based on the proposal by the SGCG was created. This template incorporates additional information regarding the classification of use cases by sub-functionalities and the affected domains and zones of the SGAM, in order to further align the specification of use cases and SGAM models and to provide the means for analyses of sub-functionalities. Further, the possibilities for the explicit description of requirements were extended and KPIs relevant to the use case can be specified within the template.

Based on the textual specification included in the M/490 working group reports, also the SGAM was formalized, resulting in the DISCERN conceptual model shown in Figure 4. One main addition provided by the DISCERN approach is the explicit model for the business layer. While the SGCG reports state possible entities located on this layer, no examples or formal semantics are provided. Also, the relations between entities on different interoperability layers were explicitly defined within DISCERN. Finally, the interrelations between use case template and SGAM were expressed in the model. This way use cases can be created to express objectives and behavioral aspects of a sub-functionality, while the SGAM models are used to provide information on corresponding architectural options.

To ensure the structural consistency of the use cases as well as SGAM models in DISCERN, templates were provided based on the findings of the conceptual meta-model. While an integrated tool-chain to support the process is planned for a later stage of the project, the first version of templates is realized based on office applications. This decision was made in order to minimize the deviation from the established toolchain at the DSOs and also to enable the intial collection of data and evaluation of the approach in an early phase of the project. While the use cases are documented in form of Microsoft Word documents, the SGAM layers as well as the valid entities on each layers are provided by a Microsoft Visio template and stencil sets. Consistency on the instance level is ensured by lists of valid functions (based on the Interface Reference Models (IRM) [4] "Abstract Components" and automation functions based on the Logical Node (LN) model defined in the IEC 61850 standard [5]), actors (based on the IRM [4], the ENTSO-E Role model [2] and the SGCG reports [12, 13, 10]) and standards (based on the SGCG reports featuring a classification within the domains and zones of the SGAM).

The approach was discussed and initially validated within a series of workshops involving the DISCERN projects stakeholders. The work of the SGCG in context of the M/490 mandate was deemed a viable basis for the project's approach, yet, as outlined in this section, several adjustments were made to extend the approach beyond the domain of standardization. The modeling approach will be further evaluated by collecting instance data on the sub-functionalities of the DSOs involved in DISCERN. This will also include validation of the infrastructure provided by templates and lists to support the creation of consistent use cases and SGAM models.

3.3 Comparison

This section will compare the experiences gained so far in context of the research projects INTEGRA and DISCERN regarding the development of a model-driven approach towards the creation of Smart Grid solution architectures. Especially the differences between the approaches shall be discussed as it is planned to evaluate the creation of an integrated approach.

Scope and Objectives

The INTEGRA project aims at a stable operation of mutually influencing Smart Grid systems. Hence, a comprehensive understanding on the single, individual systems and their interrelations should be gained. Therefore, an MDA approach should deliver a basis for reengineering and analysis of already deployed systems. The tasks of the DISCERN project are quite similar. They involve analysis of Smart Grid solutions provided by DSOs acting as the "leader" for a specific solution. The demo projects involved include both planned as well as already deployed solutions. However, the projects' main focus lies on the comparison of solutions, e.g. regarding cost-effectiveness, and also includes simulation and implementation activities. DISCERN's scope therefore may be regarded as broader in comparison to INTEGRA.

Process Model

The process models of both approaches are based on the model proposed by the SGCG. However, the process proposed in the context of INTEGRA (see Figure 2) follows the SGCG's process more strictly. DISCERN currently proposes three different options, bottom-up, top-down and combined, differing regarding the order in which the SGAM's interoperability layers are addressed.

Conceptual Meta-model

As outlined by Figures 3 and 4, the meta-models used in context of the two projects, while sharing many concepts, show some structural differences. The INTEGRA meta-model was designed with the goal of tight integration between use case descriptions and SGAM and also covers some model-transformations along the development process. Contrasting, the DISCERN model, while also outlining related concepts between both, treats use case descriptions and SGAM models as individual entities. Due to the focus on quantitative comparisons between solution models, an additional objective during the creation of the meta-model was to achieve the maximum granularity regarding the concepts included without deviation from the SGCGs source material.

Tool Support

To support the creation of use case descriptions and SGAM models, both INTEGRA and DISCERN aim at providing appropriate tool support. INTEGRA uses the SGAM-Toolbox implemented in Sparx Systems Enterprise Architect. For the reasons outlined in Section 3.2 DISCERN currently uses Microsoft Word and Visio templates based on the meta-model shown in Figure 4 to enable the elicitation of data in an early phase of the project. However, work on an integrated tool support is planned for a later stage of the project.

4   Conclusions and Future Work

This contribution was motivated by outlining the complexity regarding the development of suitable architectures for Smart Grid solutions. To achieve a thorough solution architecture, requirements have to be elicited regarding the solution's functional capabilities as well as its qualitative aspects with a special focus on interoperability. Therefore, a process to elicit these requirements is required as well as a way to express the multiple perspectives towards the proposed solutions' architectures. As domain experts provide the respective requirements it was also proposed to regard the power system domain's functional perspectives. The approach proposed to deal with these challenges is based on the results achieved by the SGCG working groups in context of the European Standardization mandate M/490, which was originally designed to support the elicitation and management of requirements in context of Smart Grid standardization by means of a use case methodology as well as the SGAM as a model to document architecture properties. This contribution extended this approach regarding three aspects:

- Raising requirements to provide formal semantics for the SGAM
- Establishing a stronger integration between the design of use cases and SGAM architecture models
- Proposing tool support for an integrated MDA-approach based on these artifacts

Regarding these aspects, experiences from two research projects, INTEGRA and DISCERN, have been considered in order to achieve a consistent solution that takes advantage of best practices gained in both projects. The next steps are concluded from the findings so far:

- Integrate the two meta-models that had been developed independently in the projects (INTEGRA and DISCERN)
- Revise the SGAM-Toolbox based on the integrated model
- Couple the UML-modeling (INTEGRA) and the Microsoft Visio visualization (DISCERN) of SGAM models to combine conceptual integrity and applicability in the working environment of the DSOs.

Acknowledgments. The research leading to these results has received funding from the European Union Seventh Framework Programme (FP7/2007- 2013) under Grant Agreement no 308913.

The financial support of the Josef Ressel Center by the Austrian Federal Ministry of Economy, Family and Youth and the Austrian National Foundation for Research, Technology and Development is gratefully acknowledged.

Funding by the Austrian Federal Ministry for Transport, Innovation and Technology and the Austrian Research Promotion Agency (FFG) under Project 838793, "INTEGRA", is gratefully acknowledged.

References

1. Andren, F., Strasser, T., Rohjans, S., Uslar, M.: Analyzing the Need for a Common Modeling Language for Smart Grid Applications. In: 11th IEEE International Conference on Industrial Informatics (2013)
2. ENTSO-E, European Federation of Energy Traders (EFET), energy Business Information eXchange (ebIX): The Harmonised Elecricity Market Role Model. Tech. rep. (2011)
3. Fang, X., Misra, S., Xue, G., Yang, D.: Smart Grid - The New and Improved Power Grid: A Survey. IEEE Communications Surveys & Tutorials 14(4), 944–980 (2011)
4. IEC: 61968-1 System Interfaces For Distribution Management - Part 1: Interface Architecture and General Recommendations (2010)
5. IEC: IEC 61850-5 Communication Networks and Systems in Substations - Part 5: Communication Requirements for Functions and Device Models. Tech. rep. (2013)
6. Locke, G., Gallagher, P.: Nist framework and roadmap for smart grid interoperability standards release 1.0
7. Lopes, A., Lezama, R., Pineda, R.: Model Based Systems Engineering for Smart Grid as Systems of Systems. Complex Adaptive Systems 1(4), 441–450 (2011)
8. Object and Reference Model Architecture Board Subcommittee: Model Driven Architecture (MDA). Tech. rep., Object Management Group (2001)
9. Object Management Group: MDA Guide Version 1.0.1. Tech. rep. (2003)
10. Smart Grid Coordination Group: First Set of Standards. Tech. Rep. November, CEN-CENELEC-ETSI (2012)
11. Smart Grid Coordination Group: Smart Grid Information Security. Tech. Rep. CEN-CENELEC-ETSI (2012)
12. Smart Grid Coordination Group: Smart Grid Reference Architecture. Tech. Rep. CEN-CENELEC-ETSI (2012)
13. Smart Grid Coordination Group: Sustainable Processes. Tech. Rep. CEN-CENELEC-ETSI (2012)
14. Trefke, J., Rohjans, S., Uslar, M., Lehnhoff, S., Nordström, L., Saleem, A.: Smart Grid Architecture Model Use Case Management in a large European Smart Grid Project. In: 4th IEEE European Innovative Smart Grid Technologies, ISGT (2013)

SP-AHP: An IT System for Collaborative Multi-criteria Decision-Making

Thomas Reichel and Gudula Rünger

Abstract. The decision-making in product and production engineering, such as the selection of an optimal design alternative, has to incorporate multiple, often conflicting decision criteria (e. g., cost, technical quality, and the environmental impact) as well as the corresponding domain experts in the decision process. Since companies are faced with competitive pressure resulting in cost and time restrictions, an appropriate IT system supporting rapid and nevertheless accurate decision processes is highly recommended. This article presents the web-based IT system SP-AHP that provides the solution to multi-criteria, multi-stakeholder decision problems. In particular, a flexible and easy-to-use approach for collaborative decision-making as well as the corresponding IT implementation is proposed. Moreover, the software architecture of the entire system and its functionalities are outlined.

1 Introduction

Design and planning decisions in product and production engineering are typically multi-criteria and multi-stakeholder processes. For each decision, multiple, often conflicting decision criteria, such as cost, technical quality, and lately the environmental impact of a product, have to be considered [12]. Due to the multiple decision criteria, multiple stakeholders of different domains, such as electrical and mechanical engineering, marketing, accounting, governmental regulations, and ecologic life cycle assessment, have to be incorporated in the decision process [9]. At the same time, companies are faced with cost and time restrictions as well as competitive pressure that require rapid and nevertheless accurate decision processes.

Thomas Reichel · Gudula Rünger
Technische Universität Chemnitz
Faculty of Computer Science,
Chair for Practical Computer Science
09107 Chemnitz, Germany
e-mail: {thomas.reichel,ruenger}@cs.tu-chemnitz.de

P.-J. Benghozi et al. (eds.), *Digital Enterprise Design & Management,*   59
Advances in Intelligent Systems and Computing 261,
DOI: 10.1007/978-3-319-04313-5_6, © Springer International Publishing Switzerland 2014

A multi-criteria decision problem consists of a set of alternatives, which are possible solutions of the decision problem, and multiple decision criteria that decompose the overall decision problem into sub-problems. The goal of multi-criteria decision-making is the selection of an optimal alternative or the prioritization (ranking) of the alternatives with respect to all decision criteria [17]. Numerous methods have been proposed that solve multi-criteria decision problems according to the preference of a single stakeholder, such as the analytic hierarchy process (AHP), the utility value analysis and the multi-attribute utility theory [3]. The incorporation of multiple stakeholders in these methods, and, hence the incorporation of different preferences, complicates the selection of the most favorable alternative.

This article presents the web-based IT system SP-AHP that supports multi-criteria multi-stakeholder decision problems based on the decision method AHP. The contribution is an approach for group decision-making in which the incorporation of stakeholders and the aggregation type of the individual assessments are defined according to the stakeholders' needs and the company restrictions. Moreover, the IT implementation of the proposed approach as well as the software architecture of SP-AHP and its functionalities are provided.

The next sections are structured as follows: Section 2 summarizes the AHP and existing methods supporting group decisions. The proposed approach for group decision-making with SP-AHP is presented in Sect. 3. Section 4 provides an example application of the approach. The IT system SP-AHP is outlined in Sect. 5. Finally, Sect. 6 concludes.

2 The Analytic Hierarchy Process

The analytic hierarchy process (AHP), initially proposed by Saaty [16], is a methodology which supports multi-criteria decisions in a widespread area of application, such as optimal alternative selection, prioritization [13], evaluation and benchmarking [4]. The first step of the AHP is the decomposition of the multi-criteria decision problem into sub-problems (i. e., the decision criteria). The decision criteria are organized in a tree structure (the so-called criteria hierarchy) with the decision problem as root node. In the subsequent assessment, the alternatives are compared pairwise with respect to each leaf node of the criteria hierarchy, followed by the pairwise comparison of (sub-)criteria with respect to their criterion above in the hierarchy. As a result, a quadratic comparison matrix is created for each criterion and the decision problem. For each comparison matrix so-called local weights are calculated that specify the assessment of alternatives with respect to a leaf criterion (resp. the assessment of sub-criteria with respect to their criterion above). The final assessments of alternatives (i. e., the global weights) are calculated using a weighted sum of the local weights. The benefits of AHP compared with other multi-criteria decision making methods is its ability to incorporate intangibles [6], its simplicity, ease of understanding, flexibility, and accuracy [4].

A key issue of AHP is its lack of scalability, because the assessment of n alternatives requires at least $n \cdot (n-1)/2$ pairwise comparisons. With that, the number of comparisons grow quadratically with the number of alternatives. Several techniques for the reduction of pairwise comparisons have been proposed (for example [13] provides an overview), but only a few are implemented in IT systems supporting the AHP. More issues of the AHP are, among others, the rank reversal problem [4], fuzzy assessments [1], and the use of alternative scales instead of Saaty's scale [8]. Methods for these issues are proposed, but most of the IT systems support only the classical AHP without any extensions [8]. In contrast, the system SP-AHP proposed in this article provides solutions for all above-mentioned issues. Therefore, the system can be applied to various multi-criteria decision problems.

If more than one stakeholder is involved in the assessment of a multi-criteria decision problem, two different problems have to be solved [2]. The first one is the so-called *selection model*. The selection model defines how to derive a final solution, e. g., an optimal alternative, from individual stakeholder preferences, i. e., the resulting comparison matrices, local or global weights. This model involves the aggregation of individual preferences as well as an evaluation of the preferences aggregated to determine the final solution. The second problem is denoted as *consensus model* [7]. It defines how different stakeholders achieve an agreement with respect to the assessment of all decision criteria. Methods for an implementation of the consensus model are based on repeated interactions between the stakeholders with or without a moderator. Herrera-Viedma et al. state in [7] that it is preferable to achieve a maximum degree of agreement between the stakeholders with the consensus model before the selection model is applied.

Dong et al. state in [2] that there are two most useful methods for implementing the selection model: *aggregation of individual judgements* (AIJ) and *aggregation of individual priorities* (AIP). The method AIJ aggregates the individual comparison matrices of a decision criterion into a final comparison matrix using the (weighted) geometric mean [11]. In contrast to the method AIJ, the method AIP aggregates the individual local or global weights of each stakeholder. For this aggregation, the arithmetic or geometric mean is used [5].

Although many methods for group decision-making with the AHP exists [2], IT systems supporting the AHP often lack the incorporation of multiple stakeholders as well as suitable consensus and selection models. The group decision algorithm proposed in this article joins the consensus and the selection model instead of a subsequent application of both models.

3   Collaborative Decision-Making with SP-AHP

The key feature of collaborative decision-making with SP-AHP is the configurable assessment of the decision criteria by multiple stakeholders and the aggregation of individual comparison matrices resulting from the assessment[1]. The input of the collaborative decision-making is the criteria hierarchy and the corresponding alternatives. The output is an aggregated comparison matrix for each decision criterion

---

[1] The collaborative identification of decision criteria and alternatives as well as the creation of a criteria hierarchy is not part of this section.

as well as the problem description. The approach for collaborative decision-making
has two subsequent steps:

1. **Configuration:** Firstly, the stakeholders specify who is responsible for the assessment of which decision criteria, i. e., the assignment of stakeholders to decision criteria. If more than one stakeholder is responsible for the assessment
   of a single criterion, the stakeholders also specify the aggregation type of the
   individual comparison matrices.
2. **Execution:** Based on the assessment configuration, the stakeholders assess the
   decision criteria assigned. According to the aggregation type of each criterion,
   a semi-automatic algorithm manages the aggregation of individual comparison
   matrices resulting in a consistent decision.

In the next subsections, both steps are presented in detail.

3.1 Assessment Configuration

The first step of the collaborative decision-making is the assignment of stakeholders.
In this step, the stakeholders specify who is responsible for the assessment of each
decision criterion. For that the stakeholders apply the following strategies.

1. **Partitioning:** In the partitioning strategy, different stakeholders compare disjoint
   sets of alternatives (resp. criteria) with each other. The strategy is primarily used
   to reduce the amount of pairwise comparisons for individual stakeholders and
   to assign domain-specific decision criteria to the corresponding domain experts.
   Due to the assessment of domain-specific decision criteria by appropriate domain
   experts, accurate results can be achieved. In the case of cross-domain decision
   criteria, the partitioning strategy may decrease the accuracy of the prioritization,
   since not all stakeholders are involved in the assessment. The partitioning strategy corresponds to the *use of incomplete experimental design* by Weiss and Rao
   [18].
2. **Consensus:** In the consensus strategy, experts of different domains assess the
   same set of decision criteria independently and then achieve a consistent assessment together. The consensus strategy is used to assess cross-domain decision
   criteria by appropriate domain experts. Hence, this strategy is primarily applied
   to consolidate the accuracy of the prioritization with respect to multiple stakeholders.

The choice between both strategies for each decision criterion primarily depends
on the decision problem and the opinions of the stakeholders itself. However, we
propose to use the consensus strategy for the assessment of decision criteria near the
root of the criteria hierarchy, since the resulting comparisons have a greater impact
on the final global weights than the assessment of lower-level criteria. For lower-
level decision criteria that may be more domain-specific, the partitioning strategy
is advisable in order to reduce the amount of pairwise comparisons for individual
stakeholders.

If more than one stakeholder assesses a decision criterion, the resulting comparison matrices have to be merged into a single matrix (the final comparison matrix). Therefore, an aggregation type has to be specified for such decision criteria. The first (and simplest) aggregation type is the geometric mean proposed by McCarthy in [11]. With this type, each entry of the final comparison matrix is calculated by the geometric mean of the corresponding entries in the comparison matrices of the stakeholders.

However, the mean value of all stakeholder assessments is not appropriate for all decision criteria. Therefore, two additional aggregation types are proposed that require an interaction between the stakeholders. However, an interaction is necessary only if the comparison matrices of the stakeholders deviate significantly. To identify such a significant deviation, an indicator is proposed for each aggregation type. The indicators are calculated for each matrix pair. We propose to use the following indicators:

- The local weights of the comparison matrices of the stakeholders result in different orders (ranks);
- The euclidean distance of the first eigenvectors of the comparison matrices (i. e., the local weight vectors) of two stakeholders is above a user-defined limit.

If the indicator of a matrix pair exceeds a user-defined limit, the stakeholders have to revise their assessments. In the manual interaction, the existing comparisons are discussed in order to achieve a high degree of agreement.

3.2 Assessment Execution

The second step is the actual assessment of the decision criteria according to the assessment configuration. Algorithm 1 defines the assessment of the decision criteria by the stakeholder specified and the achieving of consistent comparison matrices for all criteria.

First, the stakeholders assess the decision criteria assigned independent from each other (Alg. 1, lines 1–2). All stakeholder assessments start at the same time as independent processes. This concurrent assessment is defined by the term *concurrently* in line 1 as well as line 8. When the stakeholders' assignments are finished, the comparison matrices resulting are stored for the subsequent processing (line 3).

Starting with line 4, each decision criterion that was assessed by stakeholder s is examined. Line 5 checks whether all comparison matrices of criterion c are already stored by the stakeholders S_c. If all comparison matrices of criterion c are stored, it is examined whether an additional interaction between the stakeholders S_c is necessary (line 6). Such an interaction is necessary, if an aggregation type with additional interaction is specified for c and one of the indicator values for a matrix pair exceeds the user-defined limit (see Sect. 3.1).

The additional interaction of the stakeholders in the lines 7–11 starts with an invalidation of all existing comparison matrices of criterion c. With that, the existing comparison matrices are moved off the storage. Then each stakeholder in S_c revises the assessment of criterion c (line 9) and stores the comparison matrix resulting

---

Algorithm 1: Collaborative multi-criteria decision-making with SP-AHP.

---

input: C: Set of all decision criteria
input: S: Set of all stakeholders
input: C_s: Set of decision criteria that are assessed by the stakeholder $s \in S$
input: S_c: Set of stakeholders that have to assess the decision criterion $c \in C$
output: A comparison matrix for each decision criterion $c \in C$

1   **foreach** $s \in S$ **do** concurrently
2   assess all criteria C_s by stakeholder s;
3   store the comparison matrices of the criteria C_s;
4   **foreach** $c \in C_s$ **do**
5   **if** *all comparison matrices of c are stored by stakeholders S_c* **then**
6   **if** *manual interaction of stakeholders S_c is necessary for c* **then**
7   invalidate all comparison matrices of c;
8   **foreach** $s \in S_c$ **do** concurrently
9   assess the criterion c by stakeholder s;
10   store the comparison matrix of criterion c;
11   **go to** 5;
12   **else**
13   calculate the geometric mean of all comparison matrices for criterion c;
14   store the final comparison matrix of c;

(line 10). At the end of the additional interaction, the comparison matrices of the stakeholders S_c are rechecked whether one of the indicator values for a matrix pair exceeds the user-defined limit. The interactions are retried until no matrix pair exceeds the user-defined limit. However, the interactions may result in a never-ending loop if the stakeholders insist on their assessments. Therefore other termination conditions, such as the maximum number of interaction loops for each criterion, have to be integrated.

If no additional interaction is necessary or the aggregation type geometric mean is specified for criterion c, the comparison matrices are merged by calculating the geometric mean (line 13). Finally, the matrix calculated is stored in line 14.

The implementation of Alg. 1 is based on two workflows. Both workflows are depicted in Fig. 1 in BPMN[2]. Below each workflow task the corresponding lines of Alg. 1 are denoted. The so-called *assessment workflow* creates a manual task inviting a stakeholder to assess a list of decision criteria and stores the resulting comparison matrices (Alg. 1, lines 2–3 and 9–10). Each assessment workflow is executed as an independent process. When the assessment workflow terminates, the so-called *consensus workflow* is started. The consensus workflow examines the comparison matrices of the stakeholders (lines 5–6). If an additional interaction is necessary, the workflow invalidates the comparison matrices (line 7) and starts new

---

[2] Business Process Model and Notation, http://www.omg.org/spec/BPMN/

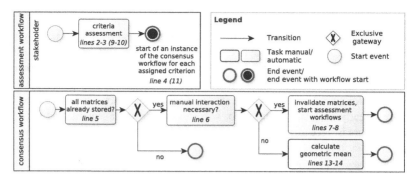

Fig. 1 Workflow implementation of Alg. 1. The lines of the of Alg. 1 which are implemented by the workflow tasks are denoted below each task node.

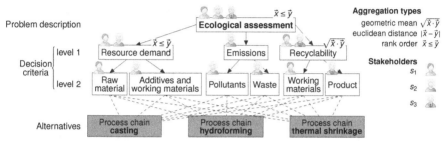

Fig. 2 Example assessment configuration for the ecological assessment of alternative process chains with three stakeholders participating. For each criterion, the stakeholders responsible for the assessment are shown in the upper left corner. The aggregation type for each criterion is shown in the upper right corner. The criteria hierarchy is an excerpt of the hierarchy proposed in [14].

instances of the assessment workflow (line 8). Moreover, the consensus workflow calculates and stores the final comparison matrix, if no additional interaction is necessary (lines 13–14).

4 Example: Assessment of Manufacturing Process Chains

As an example application, the proposed approach is applied to a criteria hierarchy for the ecological assessment of alternative process chains for the manufacturing of camshafts. Figure 2 illustrates the collaboration configuration for the assessment of process chains including three stakeholders $s_1, s_2,$ and s_3. The stakeholders responsible for the pairwise comparison of the criteria (resp. alternatives) below a criterion are shown in the upper left corner of the criterion. If more than one stakeholder is responsible, the aggregation type specified is shown in the upper right corner of each criterion.

In the assessment execution of the example given in Fig. 2, three assessment workflows are instantiated that invite the stakeholders to assess the criteria assigned. For example, stakeholder s_3 is invited to assess the criteria *Product*, *Recyclability*, and *Ecological assessment*. Once a stakeholder finishes his or her assessments, an instance of the consensus workflow is started for each criterion. The consensus workflow examines whether the assessments of all stakeholders assigned to the criterion are already performed, and if so, whether an additional interaction between the stakeholders is necessary.

For example, if stakeholder s_3 finishes the assessments, an instance of the consensus workflow starts for each of the three criteria. Assuming that s_1 has already performed his or her assessments and s_2 has not, the three consensus workflows behave as follows:

- The consensus workflow for the criterion *Product* determines that only s_3 was responsible for the criterion. Thus, the corresponding comparison matrix is stored as final assessment of the criterion *Product*.
- The consensus workflow for the criterion *Recyclability* determines that the assessments of s_1 and s_3 are already performed. Since the geometric mean is specified as aggregation type, the workflow calculates the mean values of the comparison matrices of s_1 and s_3 and finally stores the mean matrix resulting.
- The consensus workflow for the criterion *Ecological assessment* determines the lack of the assessments of s_2. Thus the workflow instance ends. As soon as s_2 finishes his or her assessments, a new instance of the consensus workflow for *Ecological assessment* determines that the assessments of s_1, s_2 and s_3 are performed. If the rank orders of the three comparison matrices are the same, the geometric mean of the comparison matrices is finally stored. Otherwise three instances of the assessment workflow are started that invite each stakeholder to revise their pairwise comparisons of the criterion *Ecological assessment*.

The assessment and consensus workflows instantiated for the stakeholders s_1 and s_2 behave in the same way.

5 Functionalities and Software Architecture of SP-AHP

The SP-AHP tool is designed to be used for arbitrary multi-criteria, multi-stakeholder decision problems. The tool is part of a software framework called EnergyNavigator [15] that provides different web applications to support engineers in developing energy efficient products. The implementation of the framework as well as the SP-AHP tool is based on the JBoss Enterprise Middleware Suite[3]. Figure 3 depicts the web-based user interface of the SP-AHP tool to be used by engineers.

The major functionalities of the SP-AHP tool are:

a. Assessment of incomplete AHP structures (i. e., criteria hierarchies and alternatives), so that only selected alternatives are pairwise compared with respect to

---

[3] http://www.jboss.org

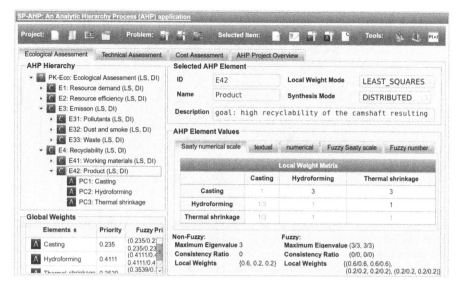

Fig. 3 Main user interface of the web-based SP-AHP software tool displaying the ecological assessment of three process chains for the manufacturing of camshafts (Fig. 2) after the assessment of all decision criteria. The properties and the comparison matrix of the selected criterion *E42: Product* are shown on the right side of the user interface. The user can change the criteria hierarchy and the comparison matrix of the selected criterion using the items of the menu bar.

the leaf criteria of the hierarchy (with that the number of pairwise comparisons can be reduced [13]);

b. Management of multiple independent AHP structures with a common set of alternatives (for example used for the separate ecological, technical, and economical assessment of manufacturing process chains, see Fig. 3);

c. Aggregation of the global weights from the independent AHP structures by a user-defined formula;

d. Support for six additional scales (Power, Root square, Geometric, Logarithmic, Inverse linear, and Balanced) outlined in [8] in addition to Saaty's scale;

e. Ideal synthesis of Saaty and multiplicative synthesis of Lootsma [10] in addition to the conventional synthesis. The ideal and multiplicative synthesis address the rank reversal issue of AHP;

f. Fuzzy specification of pairwise comparisons and corresponding calculation of fuzzy local and global weights implementing the fuzzy AHP approach of Buckley [1];

g. Support for sensitivity analyses to examine the impact of modified local weights on the resulting global weights;

h. Incorporation of multiple stakeholders implementing the partitioning and the consensus strategy of the SP-AHP by using workflows as described in Sect. 3.

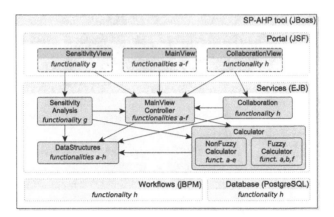

Fig. 4 Software architecture of the SP-AHP tool based on the JBoss Application Server. The architecture consists of three layers (Portal, Services, and Workflows/Database) that contain the software components of the tool. The transitions between the software components denote dependencies. The functionalities which are implemented by each component are shown below the name of the component.

The software architecture of the SP-AHP tool (Fig. 4) consists of the three layers Portal, Services, and Workflows/Database which contain the software components implementing the functionalities a–h. The layer **Portal** contains the web-based user interfaces of the tool. The *MainView* (shown in Fig. 3) is the principal interface of the tool for the creation of AHP structures, the assessment of criteria and alternatives as well as the visualization of the local and global weights calculated (functionalities a–f). The Portal also contains user interfaces for sensitivity analyses (*SensitivityView*, functionality g) and the incorporation of multiple stakeholders in the assessment (*CollaborationView*, functionality h). The user interfaces are implemented with the Java ServerFaces (JSF) framework.

The layer **Services** implements the functionalities of the SP-AHP tool. The software components *SensitivityAnalysis*, *MainViewController*, and *Collaboration* provide the functionalities for the corresponding user interfaces. The sensitivity analysis and the collaboration of stakeholders depends on the specific AHP structures provided by the *MainViewController*. Therefore, dependencies between the corresponding components and the *MainViewController* exist. The component *Collaboration* is also responsible for instantiating workflows of the assessment execution (Sect. 3.2). The software component *Calculator* implements the calculation of local and global weights as well as the corresponding consistency ratios. Since the calculations of fuzzy and non-fuzzy weights differ, the calculations are implemented as separate components. The aggregation of global weights (functionality c), the additional scales (functionality d) and the synthesis modes (functionality e) are implemented for non-fuzzy pairwise comparisons only. The component *DataStructures* contains the data structures necessary for all functionalities of the SP-AHP tool. The software components of the layer Services are implemented as enterprise java beans (EJB) and utilize the Java Matrix Package[4] for the basic linear algebra operations.

The layer **Workflows/Database** provides the workflows proposed in Sect. 3.2 and the database interface for the storage of assessments of different stakeholders. Participating stakeholders and aggregation types for the overall assessment process are chosen by a project manager (via the *CollaborationView* of the Portal). For each stakeholder, a list of assigned tasks, which contain requests for pairwise comparisons, is provided by the workflow management system. The implementation of the workflows utilizes the workflow management system jBPM and the corresponding workflow description language jPDL provided by the JBoss Suite.

6   Conclusion

The IT system SP-AHP presented in this article supports collaborative multi-criteria decisions based on the analytic hierarchy process (AHP). In contrast to existing tools for multi-criteria decision support, SP-AHP integrates several extensions of the classical AHP, which are an approach for the reduction of pairwise comparisons, fuzzy pairwise comparisons, additional judgement scales beyond Saaty's scale and techniques to avoid the rank reversal problem. With that, SP-AHP can be applied to various multi-criteria problem definitions, especially in product and production engineering. A case study is outlined in [14].

In order to support collaborative decisions, we propose a flexible and easy-to-use approach. The approach facilitate the distribution of the assessment of the decision criteria among the stakeholders and provides three aggregation types to merge the individual comparison matrices resulting. As the stakeholders itself define who is responsible for the assessment of each decision criterion and which aggregation type is used, the consensus process among the stakeholders can be adapted according to the decision problem and the company restrictions.

The benefit of the proposed approach is the ensuring of the incorporation of the preferences of each stakeholder during the assessment execution. This is especially important for large criteria hierarchies with many participating stakeholders. Moreover, the stakeholders can assess the criteria spatial and temporal independent from each other due to the workflow- and web-based implementation. However, the approach achieves a high degree of agreement between the stakeholders, and thus accurate assessment results, only with a adequate willingness to compromise.

Future work incorporates methods for the collaborative gathering and selection of decision criteria and alternatives, since a proper criteria hierarchy is important for achieving high quality assessment results. Based on further evaluations of SP-AHP in the mechanical and software engineering domain, methods for a seamless integration of SP-AHP in existing engineering tools that lack multi-criteria decision support should also be developed.

---

[4] http://math.nist.gov/javanumerics/jama/

Acknowledgements. The Cluster of Excellence "Energy-Efficient Product and Process Innovation in Production Engineering" (eniPROD®) is funded by the European Union (European Regional Development Fund) and the Free State of Saxony.

References

1. Buckley, J.: Fuzzy hierarchical analysis. Fuzzy Sets and Systems 17(3), 233–247 (1985)
2. Dong, Y., Zhang, G., Hong, W.C., Xu, Y.: Consensus models for AHP group decision making under row geometric mean prioritization method. Decision Support Systems 49(3), 281–289 (2010)
3. Figueira, J., Greco, S., Ehrogott, M. (eds.): Multiple Criteria Decision Analysis: State of the Art Surveys. Int. Series in Operations Research & Management Science, vol. 78. Springer (2005)
4. Forman, E., Gass, S.: The analytic hierarchy process - an exposition. Operations Research 49(4), 469–486 (2001)
5. Forman, E., Peniwati, K.: Aggregating individual judgments and priorities with the analytic hierarchy process. European Journal of Operational Research 108(1), 165–169 (1998)
6. Harker, P.: Incomplete pairwise comparisons in the analytic hierarchy process. Mathematical Modelling 9(11), 837–848 (1987)
7. Herrera-Viedma, E., Alonso, S., Chiclana, F., Herrera, F.: A consensus model for group decision making with incomplete fuzzy preference relations. IEEE Transactions on Fuzzy Systems 15(5), 863–877 (2007)
8. Ishizaka, A., Labib, A.: Analytic hierarchy process and expert choice: Benefits and limitations. ORInsight 22(4), 201–220 (2009)
9. Lindemann, U., Stetter, R., Viertlböck, M.: A pragmatic approach for supporting integrated product development. Journal of Integrated Design and Process Science 5(2), 39–51 (2001)
10. Lootsma, F.A.: Scale sensitivity in the multiplicative AHP and SMART. Journal of Multi-Criteria Decision Analysis 2(2), 87–110 (1993)
11. McCarthy, K.: Comment on the Analytic Delphi Method. Int. Journal of Production Economics 27(2), 135–136 (1992)
12. Niemann, J., Tichkiewitch, S., Westkämper, E. (eds.): Design of Sustainable Product Life Cycles. Springer (2009)
13. Reichel, T., Rünger, G.: Prioritization of product requirements using the Analytic Hierarchy Process. In: Proc. of the 14th Int. Conf. on Enterprise Information Systems ICEIS 2012, vol. 2, pp. 70–76. SciTePress (2012)
14. Reichel, T., Rünger, G.: Multi-criteria decision support for manufacturing process chains. Chemnitzer Informatik-Berichte CSR-13-02, TU Chemnitz (2013)
15. Rünger, G., Götze, U., Putz, M., Bierer, A., Lorenz, S., Reichel, T., Steger, D., Wenzel, K., Xu, H.: Development of energy-efficient products: Models, methods and IT support. CIRP Journal of Manufacturing Science and Technology 4(2), 216–224 (2011)
16. Saaty, T.: The Analytic Hierarchy Process. McGraw-Hill, New York (1980)
17. Vincke, P.: Multicriteria Decision-aid, 1 edn. Wiley (1992)
18. Weiss, E., Rao, V.: AHP design issues for large-scale systems. Decision Sciences 18(1), 43–57 (1987)

Blueprinting for Technology-Based Service: Decoupling of Physical-Virtual Layers

Jieun Kim and Yongtae Park[*]

Abstract. Nowadays traditional offline service processes have been significantly changed by new technology in general, and information and communication technology (ICT) in particular. That is, technology infusion has been primal way of transforming conventional offline service to advanced technology-based service. In (re)designing technology-based services, one of the most effective tools is service blueprinting, a tool to help the services maintain a customer-centered view. However, previous approaches of service blueprinting are subject to shortcoming in that they have only focused on information system-level. In order to design on-offline convergence service system, the traditional blueprint is required to provide a more holistic view, accommodating physical and virtual activities as well. To this end, this paper explores a new way of blueprinting by decoupling the 'physical' layer and 'virtual' layer. The notion of 'line of materiality' demarcates the border between two layers. The proposed approach is expected to facilitate the design and management process of technology-based service system.

Keywords: technology infusion, technology-based service, service blueprint, service system design, decoupling.

1 Introduction

The role of information and communication technology (ICT) in service sector has featured prominently in recent discussions. The traditional service organizations that once had relied on face-to-face personal contacts between customers and employees now openly adopt technology in their service delivery. Here, the strategic

Jieun Kim · Yongtae Park
Department of Industrial Engineering, Seoul National University,
1 Gwanak-ro, Gwanak-gu, Seoul, 151-742, Republic of Korea
e-mail: {hsns1234,parkyt1}@snu.ac.kr

* Corresponding author.

P.-J. Benghozi et al. (eds.), *Digital Enterprise Design & Management*,
Advances in Intelligent Systems and Computing 261,
DOI: 10.1007/978-3-319-04313-5_7, © Springer International Publishing Switzerland 2014

and operational importance of *technology infusion* arises for both customers and employees in improving the efficiency and effectiveness in service encounters (Bitner 2001). Technology infusion provides customization and flexibility, effective service recovery, and spontaneous delight (Bitner et al. 2000; Fitzsimmons and Fitzsommons 2006; Glushko 2010). Accordingly, a substantial number of services have emerged in a variety of industries, including financial services (online banking, ATMs), insurance (online consulting, online contract management), and tourism (online booking).

The extent of ICT infusion has seen a gradual increase in recent years. The advent of technological innovations enabled the convergence of offline and online service experience. Mobile and ubiquitous technology, specifically, enabled brick and mortar storefronts to become more sophisticated and better integrated with online channels. For example, a customer arriving at an offline shopping store would see his smartphone application instantaneously providing free coupons, recommendations of top sales, and the aisle location of items in need. This study focuses on effectively designing such technology-based services where offline contacts, encounters, and processes are enhanced by online technology.

When designing a technology-based service, service blueprinting has been adopted as one of the most effective tools for maintaining customer-centered view (Agnihothri et al. 2002; Fliess and Kleinaltenkamp, 2004; Bitner et al. 2008). Service blueprints can be described as two-dimensional pictures of service processes (Fliess and Kleinaltenkamp 2004). While the horizontal axis represents the chronology of actions conducted by the service customer and the service provider, the vertical axis distinguishes the areas of action. A typical service blueprint consists of five components (Bitner et al. 2008): customer actions; onstage/visible contact employee actions; backstage/invisible contact employee actions; support processes; and physical evidence. These areas of actions are separated by diverse lines: the line of interaction, the line of visibility, and the line of internal interaction.

As a whole, existing literatures have focused on designing information service systems by combining the methodologies of service management (service blueprinting) and software engineering (business process modeling) to infuse technology into service. For example, Patricio et al. (2008) proposed the service experience blueprint as a new multidisciplinary method for the design of technology-enabled multi-interface service experiences. Wreiner et al. (2009) tried to suggest service blueprints for multiple actors by separating layers according to actors. Glushko (2010) separated blueprints according to the perspectives of service system user (i.e. customer-centric and employer-centric service blueprint). Gersch et al. (2011) proposed the business process blueprinting by integrating modeling tool— i.e. architecture of integrated information systems/event-driven process chains— with service blueprinting.

However, as elucidated above, previous blueprints are subject to shortcomings in that they focus only on describing the information system or software for "lower-level" service process, where the target place for technology infusion are already decided. Thus, the previous blueprints cannot be used in making adequate decisions on the areas of technology infusion in (re)designing of offline service by on-offline convergence (Glushko 2010). Furthermore, depending on its intrinsic nature and/or functional role, technology can take on such various modes as technology-assisted, technology-facilitated, and self-service (Froehle and Roth 2004). In addition, the whole service often encompasses different modes of technology-based encounters. Therefore, choosing the specific service action to improve in service (re)designing must entail the review of "higher-level" service delivery process that properly encompasses on-offline convergence.

In the higher-level blueprint for technology-based service, both physical and virtual activities should be taken into account together in a more holistic view. That is, technology-based service involves virtual actions in addition to existing physical actions and information flows integrated with existing material flows (Morelli 2009; Glushko 2010). Consequently, in most previous higher-level blueprinting, physical activities are represented in front-office and virtual systems are noted in back-office without differentiation. However, the physical and virtual activities of service often coexist in interrelation; to account for their mutual existence, each blueprint layer incorporates the physical and virtual activities together. This approach, albeit usable, radically increases the complexity of blueprinting.

As a remedial measure, this paper suggests an alternative approach that *decouples physical and virtual zone* in service blueprinting. In service management, the decoupling of service organization-customer relationships (thus excluding the customer from the service facility's boundaries) has been focused on the maximization of service system efficiency (Chase and Tansik 1983). This study utilizes the same notion of "decoupling" in the context of materiality facilitated by technology infusion. The *line of materiality,* which separates the physical experience from the virtual, helps to design technology-based service by elucidating the following questions: which activities should be substituted by or extended to online functions?; which new activity that derives from technology infusion interacts with the activities in existing system? In order to verify the utility of decoupling, this paper develops exemplary blueprints with physical-virtual layers and explores how this separation can help design, document, and analyze a technology-based service.

The remainder of this paper is organized as follows. First, the notion of technology-based service is introduced. Second, the conceptualization and visualization of proposed approach are described, followed by an illustration case. Finally, the conclusion section offers the contributions and limitations of this research.

2   Technology-Based Service

Technology-based service is defined as 'services produced by customers for themselves, independent of direct service employee involvement, using a technological infrastructure that is provided by the service provider' (Meuter et al. 2000).

Another definition is 'services using technology, combining hardware and software' (Sandstrom et al. 2009). Infusion of technology changes the attributes of classical face-to-face service delivery (Schumann et al. 2012). First, the requirement that the customer and provider be co-located for the service provision process (Goldstein et al. 2002; Fitzsimmons and Fitzsimmons 2006) no longer

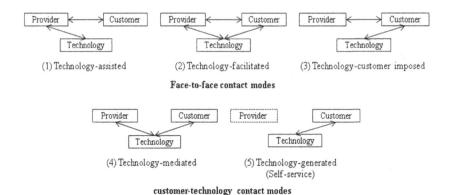

Fig. 1 Modes of technology-based service encounter (Froehle and Roth 2004; Glushko 2010)

Table 1 Characteristics of modes of technology-based service encounter (Wang et al. 2013)

Modes of technology-based service encounter	Description	Example
Technology-assisted	Technology is used by the service personnel only. However, the customer and service personnel encounter include face-to-face contact.	Hotel check-in and check-out procedures with registration system
Technology-facilitated	Although technology is used by both parties, face-to-face customer encounter still occurs.	The use of Microsoft PowerPoint by a financial expert to present and discuss financial plans with customers during a conference
Technology-customer improvised	Technology is used by the customer only. However, the customer and service personnel encounter include face-to-face contact.	Mobile application that a restaurant customer might launch on his smartphone to find wine information, and might ask the sommelier for a confirmation or second opinion
Technology-mediated	A shared techno logical platform is used by the service personnel and the customer without face-to-face contact.	The provision of professional advice by a consulting company using Videoconference technology
Technology-generated (Self-service)	Customers operate the technology without the assistance of service personnel, and face-to-screen contact replaces face-to-face customer contact.	ATM withdrawals, vending machine purchases, online shopping

holds for technology-based service exchanges. Second, technology-based services share automation, standardization and storability of service delivery. Thus, service designers infuse technology into their service delivery for customization and flexibility, effective service recovery, and spontaneous delight (Bitner et al. 2000).

Technology-based service can be classified according to the way in which technology is applied in the service production process. Froehle and Roth (2004) proposed classifying the conceptual archetypes of customer contact in relation to technology applications. This context is one of five possible modes (or distinct conceptual archetypes) of customer contact in relation to technology, as shown in Fig 1. The characteristics of each type are explained in Table 1.

3   Decoupling Physical-Virtual Layers

3.1   Conceptualization

The decoupling of physical-virtual actions in this paper is defined as 'the distinction of the channels or location of actions caused by technology infusion'. We propose the notion of *line of materiality* to separate the physical layer and the virtual layer in a service blueprint. In situations where no technology is used, face-to-face interactions engage only the physical layer. However, when technology is infused into the service system, the parts of the physical activities of the existing system are substituted or complemented by the new virtual activities. The definition of the line of materiality and physical and virtual layer are shown below.

- *The line of materiality* distinguishes the actions supported by technology and those that are not.
- *Physical layer* includes physical and tangible activities of customer and provider and is placed above the line of materiality.
- *Virtual layer* includes virtual and intangible activities of customer and provider and is placed beneath the line of materiality (area of technology infusion).

3.2   Visualization

As shown in Fig 2, the line of materiality can be drawn for every layer of service blueprint, as technology can be infused in every action of service system. The virtual zones in each layer represent the following activities:

- *Virtual zone in the customer action layer* include customer activities regarding information and/or transactions that occur through technological interface.

- *Virtual zone in onstage action layer* include information system activities that provide technological interface of service channel to customers (e.g. PC screen, mobile phone screen, and kiosk).
- *Virtual zone in backstage action layer* includes information system activities that provide technological interface of service channel to employees (e.g. Point-of-sales system screen, etc.).
- *Virtual zone in support process layer* includes backstage systems that are necessary in aiding onstage activities for delivering service.

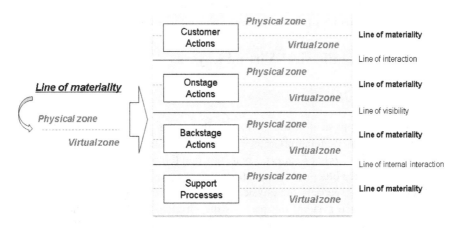

Fig. 2 Visualization of decoupling

3.3   *Illustration*

In order to explore the application of the decoupling between physical and virtual layer, we examined a service blueprint within the context of a restaurant service. A simple example of a service blueprint for a restaurant can be seen in Fig 3.

   This can be understood as technology-free service, where every process of restaurant service is conducted only by the offline service personnel and does not require the use of technology.

   We redesigned the blueprint to determine potential areas for technology infusion and appropriate mode of technology-based encounter (see Fig 4). As listed in Table 2 below, we found five examples of technology-based service encounters from Fig 3. The last column shows the areas that need virtual layers and relevant technologies for each example.

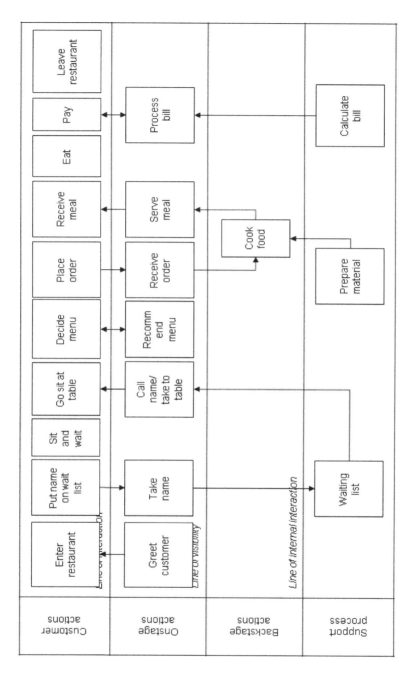

Fig. 3 Example of blueprint for a restaurant service

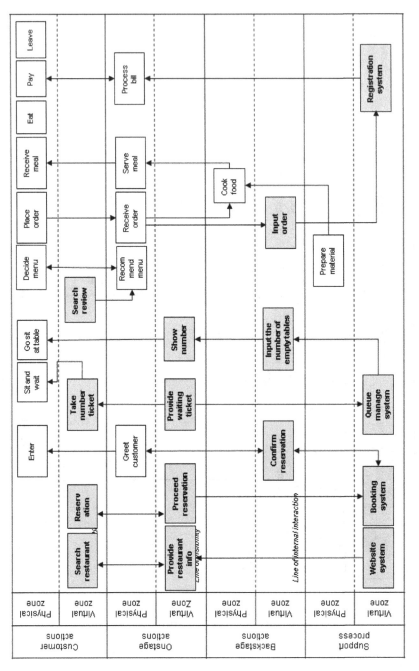

Fig. 4 Example of technology-based restaurant service

Table 2 Example of identifying areas of decoupling for technology infusion

Example of technology-based encounter of restaurant service	Modes of technology infusion	Areas that need virtual layers
Customer searches restaurant information (menu, price, location) through internet website	Technology-generated (Self-service)	Customer Onstage (internet) Support (website system)
Customer makes a reservation at the online booking system and comes to the restaurant, and employee checks the booking.	Technology-facilitated	Customer Onstage (internet) Backstage Support (booking system)
Service employee notifies the number of empty tables to the waiting system, which makes calls to waiting customers.	Technology-mediated	Customer Onstage (kiosk and screen) Support (queue management system)
Customer searches for menu reviews with his smartphone and requests employee to make recommendations.	Technology-customer improvised	Customer (mobile)
Service employee inputs order information to registration system and system provides the bill to employee.	Technology-assisted	Backstage Support (registration system)

The process of representing the above examples as in service blueprints is elaborated as follows.

- In the first scenario, the customer 'searches for restaurant information' such as menu, price, and location (the virtual action of customer). The website retrieves information from 'website system' (the virtual support process) and 'provides information' through the PC screen (the virtual onstage action). These new activities and functions emerge due to the infusion of technology.

- In the second scenario, the customer makes a 'reservation' (the virtual action of customer) in the restaurant website, which then 'proceeds the reservation' (the virtual onstage action) by requesting the 'booking system' (the virtual support process) to record the customer's information. When the customer enters the restaurant, a service employee greets customer (the physical onstage action) and 'confirms the reservation' (the virtual backstage action) through the online booking system. These new actions are also the products of technology infusion.

- In the third scenario, the customer 'takes a number ticket' (the virtual action of customer) from the kiosk that 'provides a waiting ticket' (the virtual onstage action), and the kiosk sends the addition of waiting customer to the 'queue management system' (the virtual support process). If a service employee identifies an empty table and 'inputs the number of empty tables' in the queue management system (the virtual backstage action), the system 'shows the ticket number' on the screen to call waiting customers (the virtual onstage action). This scenario shows the change of blueprint in which existing actions such as 'taking customer names' and 'including in the waiting list' are substituted by their technology-infused counterparts.

- In the fourth scenario, the customer 'searches for menu reviews' posted in blog websites with his smartphone (the virtual action of customer) and requests employees to make recommendations, and the employee promptly 'recommends the menu' (the physical onstage action). This scenario shows the change of blueprint in which existing activity of recommending menus is enhanced by the customer's utilization of ad hoc technology. In this case, the virtual action of the customer is not linked to other virtual actions because providing menu reviews is not a service provided by the restaurant.

- In the final scenario, when the customer places an order, the employee receives it and 'inputs order information' (the virtual backstage action) to the 'registration system' (the virtual support process). The system provides the bill to the employee when processing payment. By introducing the registration system, the existing 'calculating bill' activity is substituted by its technology-infused counterpart.

4 Discussion and Conclusion

This paper conceptualized the decoupling of physical-virtual layers in service blueprint and proposed its practical framework. By doing so, it is found that several advantages of decoupling may be exploited. First, decoupling of physical-virtual layers facilitates the ideation of technology infusion. Two approaches can be suggested, though several different adaptations are conceivable for varying service types. One approach is to move the existing service activities located in physical layer into the virtual layer. For example, in the third scenario, 'taking customer names' is moved to the virtual layer and is changed into 'providing a waiting ticket', prompting the idea of creating an automated system for issuing waiting tickets. Another approach is to add new activities caused by technology in the virtual layer and link them to other existing activities. For example, in the fourth scenario, 'searching menu reviews' is added because the customer unexpectedly utilizes a new mobile technology.

Second, decoupling of physical-virtual layers can help assess risks and opportunities of technology infusion in system design. Technology infusion can have both positive and negative effect on the service organization (Bitner 2001). Technology should thus be introduced not when it is merely accessible but rather when it exceeds the performance of the current offline-service constituents. Decoupling of physical-virtual layers can identify the potential problems of technology infusion by showing the interaction between the physical and virtual service elements.

Lastly, decoupling can help identify the resources in the service system. Resource is one of the three components of service system, identified by Unified Modeling Language (UML) as resources, processes, and goal. In this identification, resources are divided into the physical resource, abstract resource, and information. In line with this perspective, decoupling of physical-virtual layers can facilitate the understanding of the information flow that can be further explored in the lower-level information system development.

Last contribution of this paper is connected to the potential outlet for future study. After identifying the 'information resources' from the higher-level blueprint consisting of physical and virtual elements, one should conduct the information system design. The linkage between the higher-level blueprint suggested in this paper and the lower-level design of information system should be explored. Also, full-scale case study should be investigated in future study.

Acknowledgements. This work was supported by the National Research Foundation of Korea (NRF) grant funded by the Korea government (MEST) (No. 2011-0012759).

References

Agnihothri, S., Sivasubramaniam, N., Simmons, D.: Leveraging technology to improve field service. Int. J. Serv. Ind. Manag. 13(1), 47–68 (2002)

Bitner, M.J.: Service and technology: opportunities and paradoxes. Manag. Serv. Qual. 11(6), 375–379 (2001)

Bitner, M.J., Brown, S.W., Meuter, M.L.: Technology infusion in service encounters. J. Acad. Mark. Sci. 28(1), 139–149 (2000), doi:10.1177/0092070300281013

Bitner, M.J., Ostrom, A., Morgan, F.: Service blueprinting: a practical technique for service innovation. Calif. Manage. Rev. 50(3), 66–94 (2008)

Chase, R.B., Tansik, D.A.: The customer contact model for organization design. Manag. Sci. 29(9), 1037–1050 (1983)

Fitzsimmons, J.A., Fitzsimmons, M.J.: Service management. McGraw Hill, New York (2006)

Fliess, S., Kleinaltenkamp, M.: Blueprinting the service company – managing service processes efficiently. J. Bus. Res. 57(3), 392–404 (2004)

Froehle, C.M., Roth, A.V.: New measurement scales for evaluating perceptions of the technology-mediated customer service experience. J. Oper. Manag. 22, 1–21 (2004)

Glushko, R.J.: Seven contexts for service system design. In: Maglio, P.P., Kieliszewski, C.A., Spohrer, J.C. (eds.) Handbook of Service Science, Service Science: Research and Innovations in the Service Economy (2010), doi:10.1007/978-1-4419-1628-0_11,219249

Glushko, R.J., Tabas, L.: Designing service systems by bridging the "front stage" and "back stage". Inf. Syst. E-Bus Manage. 7, 407–427 (2009)

Goldstein, S.M., Johnston, R., Duffy, J., Rao, J.: The service concept: the missing link in service design research? J. Oper. Manag. 20(2), 121–134 (2002)

Morelli, N.: Service as value co-production: reframing the service design process. J. Manuf. Tech. Manag. 20, 568–590 (2009)

Patrício, L., Fisk, R.P., e Cunha, J.: Designing multi-interface service experiences: the service experience blueprint. J. of Service Res. 10, 318–334 (2008)

Sandstrom, S., Edvardsson, B., Kristensson, P., Magnusson, P.: Value in use through service experience. Manag. Serv. Qual. 18(2), 112–126 (2008)

Schumann, J.H., Wünderlich, N.V., Wangenheim, F.: Technology mediation in service delivery: A new typology and an agenda for managers and academics. Technovation 32(2), 133–143 (2012)

Wang, C.C., Chen, M.C., Hsien, T.C.: An Investigation of the correlation among the technology mode, service evidence and service quality from a customer perspective. Open. J. Bus. Manag. 1, 45–53 (2013)

Wilson, A., Zeithaml, V., Bitner, M.J., Gremler, D.: Services marketing – integrating customer focus across the firm, 1st European edn. McGraw-Hill, London (2008)

Urbanization of Information Systems: An Outdated Method?

Emmanuel Bertin and Noel Crespi

Abstract. The urbanization of Information Systems has been a widely employed method for years. But changes in the organization of enterprises are currently challenging it. Moreover, some of its original assumptions derived from the urbanization of cities are also questioned. In this context, this article exposes how the urbanization can be renewed by inspiring from the new city planning trends, and move from a function-based urbanization to a flow-based urbanization.

1   Introduction

The organization of enterprises is changing. The functional structure as promoted by Taylor and Ford at the beginning of the previous century is less and less useful in a fast moving environment, where firms have to compete at the innovation frontier- whatever technological, marketing or value chain innovation. In this new context, what become the principles of Information System (IS) urbanization, as recommended during the 90ies?

This paper is investigating this question, by revisiting the initial assumptions of the IS urbanization. In the first section, we remind these initial assumptions and their impact on the IS, but also their underlying source in the city planning. In a second section, we prolong the usual parallel between cities and IS by presenting the recent trends in city planning, as well as possible analogies with the IS. In the third and last section, we proposes a new method to move from a function-based to a flow-based urbanization.

Emmanuel Bertin
Orange Labs – 42, rue des Coutures
14000 Caen, France

Noel Crespi
Institut Mines-Telecom,
Telecom SudParis – 91000 Evry, France

P.-J. Benghozi et al. (eds.), *Digital Enterprise Design & Management*,
Advances in Intelligent Systems and Computing 261,
DOI: 10.1007/978-3-319-04313-5_8, © Springer International Publishing Switzerland 2014

2   The Assumptions of Urbanization

2.1   *Urbanization, in a Nutshell*

Urbanization mainly consists in three main activities:

- cartography of the existing IS
- conception of a target IS
- determination of a migration path from the existing to the target IS

In other words, according to [1]:

> "Urbanization is to organize the gradual and continuous transformation of
> information system to simplify it, to optimize its added value and to make it
> more responsive and flexible towards strategic business changes, while
> relying on technical opportunities of the market. Urban planning defines
> rules and a coherent, stable and modular context, in which different
> stakeholders are referring to any investment decision in the Information
> System."

To achieve these goals, the IS is usually represented with four viewpoints:

- the business viewpoint is composed by the processes describing the activities of
  the whole enterprise (for example fulfillment, assurance or delivery processes)
- the functional viewpoint is composed by the functions supported by the IS, and
  that are necessary for the realization of the business processes (and thus for the
  realization of the enterprise's activities)
- the applicative viewpoint is composed by the software applications of the IS
- the technical viewpoint is composed by the infrastructural components supporting the IS software applications.

To sum-up, the business view responds to the question of 'why', the functional
view responds to the question of 'what', the technical view responds to the question of 'with what' and the applicative view responds to the question of 'how' [2].

In the reference methodologies, like in the seminal work from Christophe
Longépé [3], the most operational viewpoints are usually the functional one and
the applicative one. The applicative viewpoint enables indeed to make a detailed
description of the existing software programs and data from the IS, and to propose
a target. The functional viewpoint is the key tool in order to conceive this target.
Indeed, the identification of the main IS functions enables then to make a blueprint
map of the IS, in order to position the various applications on it. This leads finally
to the definition of a target reference model, where a function is usually implemented by one and only one software application (in order to prevent a function to
be scattered between various applications).

Functions are thus the key concept of urbanizations methods. Moreover, functions are usually split with a hierarchical division: functional zones are divided

into functional districts that are themselves divided into single functions. The functional coherence should be high within a subdivision: a district groups highly coherent functions, and a zone groups highly coherent districts. Implicitly, these fundaments rely on a city planning paradigm.

2.2   The Metaphor of the City

Usually, city planning is indeed grounded on geographical invariants. A city is divided into zones, quarters and blocks. City planning rules are then enacted and enforced for each area. Obviously, these divisions must be stable in the time, and the stakeholder of the city can therefore consider them as invariant. As detailed in [4], the urbanization of the IS relies similarly on the supposition that stable functional blocks can be identified, if not forever at least for a long time. In his first urbanization work for the Société Générale, Jacques Sassoon [5] considered indeed banking products as the basic units offered by a bank. Similarly, Christophe Longépé [3] defines six functional zones (an exchange zone, a data repository zone, a referential zone for data and rules, a steering zone, an operations zone, a resource zone) in his work.

Let us now discuss how these city planning principles have been applied in real-world towns, and with which consequences. The principles from the Functional City have been defined in 1933 par Le Corbusier, with his famous "Athens charter". The ideal city is composed of blocks, where one block is dedicated to one and only one function. Four functions were identified: to live, to work, to entertain and to move. Human flows can be easily predicted (e.g. from the living block to the working block each morning), and the transportation infrastructures are conceived accordingly. The functional zoning enables to plan minimal dependencies between zones.

These theoretical principles have been massively applied to rebuild European towns after the Second World War. The governance of this urbanization process has been characterized by a tight centralization, driven by a strong administrative power (incarnated in France by senior civil servants, like the "ingénieurs des Ponts et Chaussées"). Key decisions were taken at a nation-wide level, without a real adaptation to each territory. The same types of buildings with the same framework have been for example built in Lille, Toulouse or Nancy. A strict regulation has led to a standardization of tools and methods (e.g. masterplans, blueprints). Inhabitants were not considered as stakeholders, but as resources that should remain stable and accountable.

We have to admit this functional paradigm has been produced mixed results, to say the least. With economical changes, work has disappeared from the working areas, and people from the living areas have suffered from mass unemployment, as the transportation flows have not been design to connect them to other opportunities.

2.3   The Functional Model for the IS, Some Consequences

Not only have the main urbanization methods followed the Functional paradigm, but the ERP tools also have. For example, figure 1 indicates some SAP modules that can be deployed in a firm. A Functional cutting can be easily noticed.

Within this paradigm, the governance of the IS tends to be centralized, as for cities. Ex-ante validation committees are involved to enforce the rules, and each IS project should be so validated. Transversal enterprise architects are dedicated to plan the evolution of the IS, and are natural members of the validation committees. Chief architects are here in a most powerful position than project owners. Moreover, the paradigm is usually linked with a Functional organization of the firm, which is also divided into functions, like commercial function, production, R&D, finance...

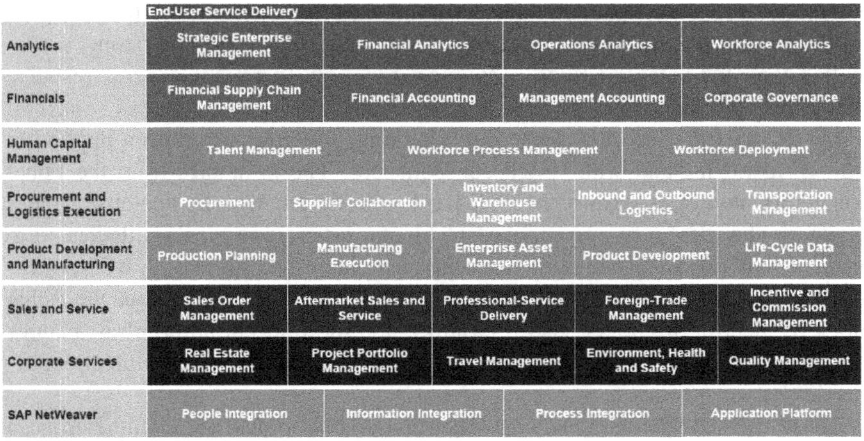

End-User Service Delivery				
Analytics	Strategic Enterprise Management	Financial Analytics	Operations Analytics	Workforce Analytics
Financials	Financial Supply Chain Management	Financial Accounting	Management Accounting	Corporate Governance
Human Capital Management	Talent Management	Workforce Process Management		Workforce Deployment
Procurement and Logistics Execution	Procurement	Supplier Collaboration	Inventory and Warehouse Management	Inbound and Outbound Logistics	Transportation Management
Product Development and Manufacturing	Production Planning	Manufacturing Execution	Enterprise Asset Management	Product Development	Life-Cycle Data Management
Sales and Service	Sales Order Management	Aftermarket Sales and Service	Professional-Service Delivery	Foreign-Trade Management	Incentive and Commission Management
Corporate Services	Real Estate Management	Project Portfolio Management	Travel Management	Environment, Health and Safety	Quality Management
SAP NetWeaver	People Integration	Information Integration	Process Integration	Application Platform

Fig. 1 Example of SAP modules

However, like for the cities, the strict enforcement of a Functional paradigm has obviously some drawbacks:

- First, the business of the firm can be rapidly evolving due to commercial, economical or technological disruptions. If this was still not fully the case in the 90ies, we now live at a time of uncertainty and unpredictability. Identifying invariant rules becomes a challenge. Enforcing they strictly might become a handicap in the worldwide competition.
- Second, a Functional model is hardly adaptable. A functional model structured by invariants will typically evolve less rapidly than the business environment. Innovative applications can be positioned on the existing functions only with difficulties or contortions.

- Third, the Functional paradigm tend to a coupling between the IS and the organization of the firm. In enterprises following this urbanization paradigm, the organizational structure is often inspired by the division in functional zones. This can slow down the decision making, due to necessary trade-offs between functions.
- Fourth, the end-user is not directly taken into account in the Functional paradigm, whatever internal end-users (i.e. employees) or external ones (i.e. clients). A single end-user need will typically necessitate the execution of many applications, from various functional zones, to be fulfilled. However, dependencies between applications are not directly represented on the Functional map, as well as potential incompatibilities. An end-to-end vision of the fulfillment of end-user requests is sometimes lacking.

The limitations of the Functional paradigm finally seem more or less similar concerning the cities and the IS. This statement encourages us to review the topical urbanization paradigm of the nowadays city.

3   A New Urbanization Paradigm, The Multipolar City

Concerning city planning, the Functional paradigm has outlived its time. A city is no more conceived as a juxtaposition of different functional zones, but as a hub to enable the circulation of flows. The idea is to recreate some continuity between the various functions. Each area is intended to be multifunctional, by grouping for example housing, offices and commercial or entertainment activities. The aim is to attract the people that are also multidimensional: working in transportation, doing sport in the middle of their working day, or shopping just in front of their office after work. The gist of the multipolar paradigm can be summed up in one sentence: flows should prevail over premises. The hub and spoke paradigm is thus a key concept: as well financial then human flows are running from hubs to hubs, without stopping at middle places. A new urban fabric will emerge around these hubs, mixing offices, housing and multiplex entertainment centers. Transportation infrastructures become then a key tool for the Multipolar paradigm: for example roads, trains or planes between cities, the tramway for suburbs. Figure 2 illustrates this by representing the connections between main French cities.

The governance is now decentralized, but with an indirect centralized but remote control, for example via founding agencies promoting a common vision through calls for project proposal, or via the financing of infrastructures (e.g. transportation infrastructures, but also museums, public centers, etc.). The aim is to empower the actors of the town (local and regional elected officials, architect...), while guiding them toward a centrally-decided direction.

We believe that this new urban paradigm can also serve as a metaphor for the IS. This would mean multi-functional zoning, ex-post validation, tools to manage the intensification of the flows, etc. In particular, transposing this new paradigm to the IS would mean to conceive the IS as a system to manage flows, instead than a system to manage data. The key stakes are thus the following:

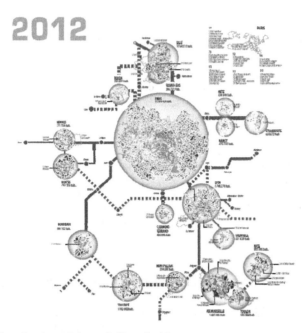

Fig. 2 Hub and spoke view of the main French cities

- The delivery of the flows, how to orient them, how to manage access rights, how to communicate between applications?
- How to represent the flows in an end-to-end way and to control them?
- How to take into account the end-user, as a source of flow as well as an actor to treat them?
- How to more dynamically adapt the IS to the business evolutions?

This model is still not generally applied, but it echoes issues currently raised by enterprises [6]. It is also aligned with the recent proposals to modify enterprise architecture concepts, like in the works from Pierre Pezziardi [7].

4   From a Function-Based to a Flow-Based Urbanization

Let us now study how this flow-based urbanization could be achieved. The foundations would be the following ones.

First, business processes are a key element, underexploited in the classical Functional paradigm.

- They provide a dynamic view of the enterprise business.
- They can easily adapt to the changes in the business activities.

- The end-user can be integrated as an actor, that not only triggers activities but that can also be involved in their realization.
- They enable to identify the business data manipulated by the different actors from the IS
- Finally, they are easily shareable within the enterprise (architects, business owners, employees…)

Second, technological patterns enable to represent the infrastructure that supports the flows in the IS.

- They present a relational view of the technological solutions used in the enterprise
- They are independent of the realized functions
- They enable to identify hubs and spokes

Two technological patterns are for example represented in figures 3 and figure 4: the IMS (IP Multimedia Subsystem) pattern – coming from the telecom world, and the SOA pattern (Service Oriented Architecture) – coming from the IT world.

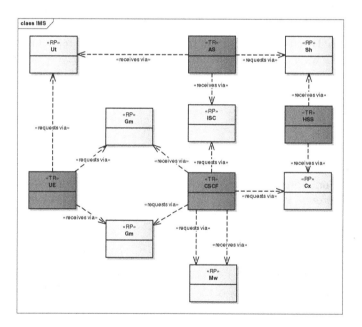

Fig. 3 The IMS technological pattern

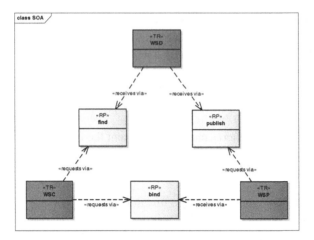

Fig. 4 The SOA technological pattern

Third, a new correlation between the business and the technological view is still to invent.

• To represent how each business need give rise to applicative flows
• To represent how each application is transforming and retargeting them

This flow-based urbanization relies on the distinction between reference views and architecture views. Reference views are stable during the lifecycle of an application, whether service views are dedicated to a given project (c.f. figure 5).

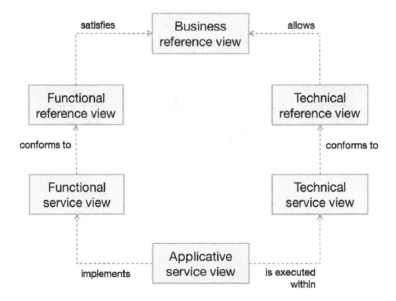

Fig. 5 Links between reference views and service views

The reference business view describes the business processes of a firm. The functional reference view formalizes the information coming from the business processes. The technical view describes the technological patterns used in the firm.

The functional service view formalizes the functional perimeter of a given project. The service technical view describes what technical components are used within this project. And the service applicative view details how the information flows (coming from the service functional view) are running in an end-to-end manner through the components coming from the technical view.

5   Conclusion

Enterprises are now living in a fast changing business environment, subject to commercial, financial and technical evolutions that can occurs in only one or two years. Agility becomes thus a cardinal virtue. The organization of the firm should be adapted to this, and so does the urbanization of the IS. The new urban Multipolar paradigm provides an interesting metaphor to help us moving from a function-based static urbanization to a flow-based dynamic one.

The key stake is here to take into account the various flows passing trough the IS – to understand them by describing them, in order to still master the end-to-end functioning of the IS at a time of uncertainty. The principles introduced in this article are further detailed in [8] and [9].

References

[1] http://www.urba-ea.org/
[2] Simonin, J., Bertin, E., Le Traon, Y., Jezequel, J.M., Crespi, N.: Business and information system alignment: a formal solution for telecom services. In: Proceedings of ICSEA 2010: The Fifth International Conference on Software Engineering Advances, CPS, Nice, France, August 22-27, pp. 278–283 (2010)
[3] Longepe, C.: The Enterprise Architecture IT Project: The Urbanisation Paradigm. But-terworth-Heinemann, Oxford (2003)
[4] Contini, I.: L'apport historique de l'urbanisme des villes pour l'urbanisme des systèmes d'information. Colloque Urbanisation des Systèmes d'Information, Paris (2002)
[5] Sassoon, J.: Urbanisation des systèmes d'information, Hermès, Paris (1998)
[6] Bidan, M., Rowe, F., et Truex, D.: An empirical study of IS architectures in French SMEs: Integration approaches. European Journal of Information Systems 21, 287–302 (2012)
[7] Pezziardi, P.: Lean Management: Mieux, plus vite, avec les mêmes personnes, Ey-rolles, Paris (2010)
[8] Bertin, E., Crespi, N.: Architecture et gouvernance des services de communication, Hermès Lavoisier, Paris (2013)
[9] Bertin, E., Crespi, N.: Architecture and Governance for Communication Services. Wi-ley ISTE, London (2013)

Improving Sustainability of Energy Production: Design of a Demand Side Management System Using an Auction Model

Vincenzo Di Lecce, Michele Dassisti, Domenico Soldo, Antonio Giove,
Alessandro Quarto, Angelo Galiano, and Michela Chimienti

Abstract. This work presents an intelligent demand side management (DSM)
system modeled according to an auction based multi-agent system (MAS). The
system is designed to improve the sustainability of energy self-production systems
thanks to energy saving features while guaranteeing the maintenance of the user's
desired comfort level. The proposed system is composed of a sensor network and
a central processing unit. Each network node is handled by an agent and it is able
to regulate the power consumption of a single environment (e.g., a room). The
first live tests were carried out within a public building. Results seems promising
for maximizing the sustainability as well as the profitability of self-production
energy systems.

Vincenzo Di Lecce
DEI, Politecnico di Bari, Via Orabona 4, 70126 Bari, Italy
e-mail: v.dilecce@aeflab.net

Michele Dassisti
DMMM, Politecnico di Bari, Viale Japigia 182, 70126 Bari, Italy
e-mail: m.dassisti@poliba.it

Domenico Soldo · Antonio Giove
AEFLAB, Politecnico di Bari, Viale del Turismo 8, 74123 Taranto, Italy
e-mail: {d.soldo,a.giove}@aeflab.net

Domenico Soldo · Alessandro Quarto
myHermes Srl, Corso Italia 63, 74121 Taranto, Italy
e-mail: alessandro.quarto@myhermessrl.com

Antonio Giove · Alessandro Quarto · Angelo Galiano
Dyrecta Lab Srl, Via V. Simplicio 45, 70014 Conversano (BA), Italy
e-mail: maurizio.galiano@dyrecta.com

Michele Dassisti · Michela Chimienti
Laboratorio Kad3 Scarl, Contrada Baione snc, 70043 Monopoli (BA), Italy
e-mail: m.chimienti@laboratoriokad3.com

P.-J. Benghozi et al. (eds.), *Digital Enterprise Design & Management*,   93
Advances in Intelligent Systems and Computing 261,
DOI: 10.1007/978-3-319-04313-5_9, © Springer International Publishing Switzerland 2014

1   Introduction

The new "ecological sensitivity" of the last decades pervades each sensible aspect of our lives. Sustainability, the new key word of our societies, refers to the problem of minimizing resource consumption while maximizing satisfaction of human requirements. If left unattended or poorly handled, energy issues could cause problems for future business activities and occupy management's time for decades to come. Just thinking about the companies that, being heavy energy consumers, create major greenhouse gas emissions through their operations, or those that manufacture energy-dependent products; it must be clear how important is to address this issue by searching for new products able to assure energy quality standards and keep costs down at the same time. From this point of view, the optimal control of energy production and consumption, at any level of the present human activities, is a key element to improve the sustainability.

This paper presents the first outcomes of a project aiming at defining an effective soft technology for monitoring and managing power consumptions based on a self-regulation (say "intelligent") strategy; this characteristic was obtained through the agent technology, a fundamental paradigm of artificial intelligence.

Within this paper, the proposed multi-agent system technology was implemented in a control system for the management of power consumption of a public building (a University Faculty) equipped with photovoltaic panels. The new renewable energy production systems have brought the Italian power providers to define with the users new kinds of contracts such as the one allowing the user to "sell" the produced energy to the same power provider.

The paper is organized as follow: section 2 introduces the reference case; section 3 describes the MAS focusing on the proposed auction based model while section 4 reports the experiments and results. Finally, conclusions and final remarks are reported in section 5.

2   The Case Study

For the present application it was decided to test the MAS solution adopted to be implemented in a public building hosting an University faculty in southern Italy equipped also with renewable energy sources (solar panels), and in possible coordination with other forms of energy supply such as the thermal one (solar thermal collectors). This kind of application presents strong management criticalities: building is irregular and variously predictable, depending on the activities carried out and on the personnel working time. The preexistence of electrical and thermal plants, allows to evaluate the performance of the proposed system. This implies, consequently, an estimation of the critical zones of operation very far from the reference condition, where a simple controller Proportional-Integral-Derivative (PID) may ensure good regulation.

The hypothesis is that the Italian power provider offers the user the possibility to sell a quote of the produced power. In this case, the limit of users' power consumption has an economical nature: having a power consumption smaller than a given threshold means having an increase of the power sold.

At a first analysis, considering the University building energy requirements, it is clear that the major power consumption is due to the air conditioning system. In particular, each room has its own air conditioner and other fixed loads such as computers, printers, lights, etc.

As a working constraint, the auction system has been tuned to manage the power consumption of the air conditioner for each room. This constraint does not affect the generality of the proposed approach, but it is useful to define the sensors and the relays to be installed at each sensor node.

The multi agent-based energy management system implemented is aimed at monitoring and controlling complex power systems. The proposed approach acts as a multi-objective scheduler, having the following objectives:

- minimizing the power consumption;
- giving maximum comfort to user, namely satisfying as many requests as possible, by spending the minimum time (and/or money).

These objectives are achieved through the competition of more agents [1]. Each agent manages a room of the building. The proactivity features of agents and their communication ability give the proposed system a great scalability potential. Each user (owning a power production plant) can sell the provider the difference between the power produced and consumed. This fact translates a power consumption optimization into a monetary gain for the user and thus, the research for smart energy management systems is having a new impulse. The proposed agency uses the set of distributed measurements performed in each facility room as if it was a knowledge base.

This DSM system could be also adopted in large business facilities, as well as machine halls, and wherever electrical devices are used with the presence of a power production plant.

The proposed system was modeled using a simulation realized in Matlab® code. For testing the performance of the proposed system, also without using the data measured in the field, the authors have defined a model simulating the thermal behavior of a building. This model considers the following aspects affecting the building thermal: 1) seasonal thermal variations; 2) thermal behavior of the building envelope; 3) thermal behavior of the partition walls; 4) daily variation of the people's flow in the various rooms; 5) the impact of the air conditioner on the thermal conditions of the rooms.

For the sake of brevity, since data are easily found and does not bring new hints to the discussion, these aspects will not be reported.

3   MAS Architecture for DSM Application

A lot of different software architectures are presented in literature. Many of them are based on the use of a specific communication language. Other authors produce high level contributions about the structural approach of internal agent organization. The innovation proposed in the approach proposed, according to [2,3], are the interaction rules of the multi agent system architectures, where agents belongs to three macro classes:

- Interface area: the agent translates the message from external source (also human/natural language) into an Agent Communication Language (ACL). Its goal is to interpret external queries according to services offered by the MAS.
- Brokerage area: it analyzes a local database where services offered by MAS are stored. In this way, starting from one query, it produces as many messages as the request needs. It aims at managing internal information traffic flows.
- Analysis area: this group of agents attempts to validate data, using the knowledge of the system regarding their own interpretability.

Five sub-classes are introduced to enhance system scalability: Interface (one agent), Broker (one agent), Validation (more agents), Forecast (more agents) and Coach (one agent). The innovation presented in the application concerns the management strategy to allow the architecture to "intelligently" behave, thus devising appropriate interaction rules to let the system perform coherently with the customer's needs as well as to be consistent with unpredictable changes in energy demands.

Taking the faculty application as reference, here the design and implementation solutions are explained in detail. For the DSM application in object, the following agent areas were defined. A schematic overview of the proposed MAS is reported in Figure 1.

- **User Area.** In this layer there is the association between each monitored room and a corresponding "Room Agent". The Room Agent is the interface between a room and the auction agency. The main task of this agent is to manage an acquisition node from the real world allowing to evaluate environmental conditions and apply the MAS rules by acting on a relay. Each Room Agent produces a detailed profiling of users' habits in order to define a user's behavioral model. Each acquisition node is equipped with a dedicated sensor set. The specific sensor set composition is defined following the room characteristics, such as: main destination, dimensions and plant, furniture, number of doors and windows. Outputs required for each sensor set are:

  – **Environmental Information**: this kind of information is provided by a multi-source environmental monitoring system for indoor workplace. The environmental conditions are monitored in order to define the necessity or not to provide energy to the room (e.g., turning on/off the air conditioner).

The main components of the proposed Environmental Condition Detector are: a temperature evaluation block; a humidity evaluation block; an air quality monitoring block. Data collected from the previous blocks can be processed to obtain indirect information, such as the number of workers being contemporarily present as consequence of air quality degradation and/or temperature variation [4].

–   **Presence Information**: the systems monitor the human presence in order to avoid energy consumption in case of long worker absence. Starting from these data, a user behavior analysis for profiling activities is realized. The provider of presence information can be a commercial presence detector. The detector should be characterized by a square configurable detection area, so as to avoid the risk of unrevealed workers in particular conditions like the room corners.

–   **Energy Consumption Information**: these data are acquired in reference of real-time power consumption. The detector device adopted should be based on different monitoring levels. In particular, for a DSM aimed at air conditioning system regulation, it is possible to acquire information about: power consumption of each air conditioner installed in the room, power consumption of the whole room, power consumption of the area where the room is. The Energy Consumption Detector is based on a commercial energy monitoring system.

• **Interface Area.** This area is designed to allow the communication between the agents in the User Area and those in the Analysis Area. In particular, this layer is responsible for the evaluation of information produced by Room Agents in order to extract additional information (e.g., user presence probability, unease level perceived by user in function of air conditioning necessity). Evaluations realized by Agents operating in the Interface Area produce the auction bids.

• **Analysis Area.** The Analysis Area hosts two kinds of agent: the Auction Manager and the Power Manager Agent. The Auction Manager, responsible for the auction management, acquires and evaluates each bid that Interface Agents produce. In particular, the Auction Manager verifies that bids are proposed with respect to specific and predefined bidding rules. For each bidding round the Auction Manager communicates the winners to Power Manager Agent and collects money from winners. The Power Manager Agent defines power supply available for bidding round and authorizes winning Room Agents to acquire required power supply in function of air conditioning systems.

• **Knowledge Area.** The Knowledge Area is responsible for the storage of historical data acquired during system activity. The stored logs are related to: electrical consumption, environmental condition, and user presence. This knowledge base permits to deduce additional information such as: typical working hours and typical working day for each monitored room.

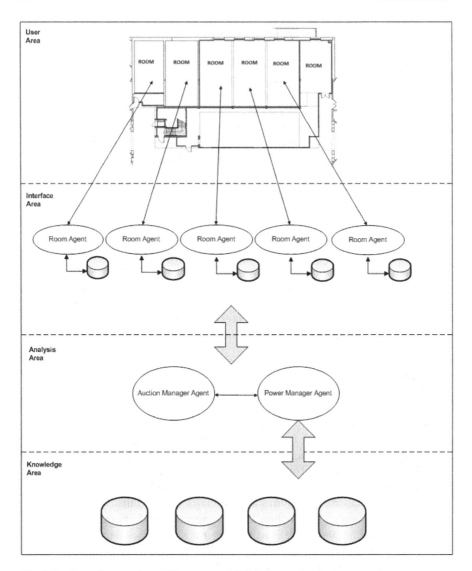

Fig. 1 A schematic overview of the proposed MAS

3.1   *Implementation of the Proposed Auction Model*

The kernel of the proposed power management system is an auction based MAS. The cooperation between the agents described in the previous sections, is aimed at defining of an auction model. This means that the whole agency behavior is organized according to specified rules. Rules are short declarations finalized to define each agent behavioral model. A rule is composed of two parts: a conditional clause and a declarative part [5].

The process rules are defined in relation to different aspects of the negotiation to manage the auction: starting auction rules, bidding submission rules, and auction management rules. A detailed list of these rules is reported in Table 1.

Table 1 Auction based MAS behavior rules

Starting auction rules

1 If a room exists then a Room Agent exists

2 If the last auction was realized G hours before then there would be an auction round

3 If a Room Agent participates to an auction round then this pays F coins

4 If a new day of biddings is instantiated then each Room Agent holds T coins

5 For each Room Agent there is a wallet W containing coins available for bidding

6 If a new auction round is realized then a global amount of power supply H is available

7 If a new auction round is realized then H is completely assignable

8 For each Room a required power supply Q for Air Conditioning system exists

Bidding submission rules

9 For each Room and each auction round there is a presence probability PP – range [0;1]

10 For each Room and each auction round there is an unease level UL – range [0;1]

11 For each Room and each auction round there is a Room Agent's bid B of U coins where B = f(UL,PP,W) – range [0; W]

12 If a Room has PP = 0 then the Room makes B = 0

13 If a Room has UL = 0 then the Room makes B = 0

Auction Management Rules

14 For each B an Auction Manager evaluation exists

15 For each auction round the Auction Manager produces a standing ordered by B's amount (starting from bigger B to lower)

16 For each auction round a set of winning Rooms exists

17 If a "Room Agent 1" bids U1 coins and a "Room Agent 2" bids U2 coins where U1>U2 then the "Room Agent 1" has priority on power supply than "Room Agent 2"

18 If a winning "Room Agent" requires a power supply Q then the residual power supply available is delta H=H-Q

19 If a winning "Room Agent" requires a power supply Q > deltaH then the bid is considered NULL

20 If a Room Agent wins an auction round then this pays the bid coins

21 For each auction round the Auction Manager acquires a jackpot J = SUM (winners's B)

22 For each auction round a Room Agent receives (J/number of Room Agents) coins

23 If a Room Agent wins an auction round and requires Q<=deltaH then the Power Manager Agent assigns him Q

As shown in Table 1, the main MAS target is the minimization of daily power consumption. For this target the system organizes a new auction instance each G hours, where G is a configurable parameter. The minimal auction round granularity is set to one hour in order to consider a minimal lot of time for guarantee user's perception.

During auctions, each Room Agent has a personal budget (the wallet W in rules of Table 1). Each Room Agent has an attitude to offer virtual money directly proportional to the user requirements. For example, if an agent has revealed a comfortable temperature in the room, it will enter in the auction with a low bid and vice- versa. The condition where an agent finishes too early its budget raises an alarm for the system manager. Indeed, in normal conditions, an air conditioner should be able to reach a stable condition in the room in a reasonable time. If this does not happen it is possible that the air conditioner is not well sized for that room or that there is an inappropriate user's behavior (e.g., the air conditioner is on while the door or windows are opened).

The agency is organized on stable amount of global money. In particular, for each auction round, money acquired by Auction Manager, is equally distributed to all the Room Agents. This condition allows for the persistent participation of any Room Agent to any auction round. So it is possible to get a greater success chance for those bidders that had a conservative approach in previous auction round.

At the same time, it is important to avoid that a Room Agent corresponding to a long time empty room collects the whole money available in the agency. This is realized with a daily auction initialization (Rules 4 in Table 1). This permits to make bids for any Room Agent starting from a common daily condition.

4   Experimental Live Tests

In order to show the functionality of the system a set of tests were performed. During these tests all the possible (real) combinations of outdoor and indoor parameters are used. The mean values of the results obtained are reported in this section. In order to highlight the impact of the proposed power management system on the power consumption optimization and on the level of guaranteed wellness, the system was tested with and without DSM. In particular, the building thermal behavior and the level of wellness achievable inside are here presented for the system with the air conditioner plant.

The results obtained for two reference months (January and July) are reported. These two months are the most expensive from an energetic point of view because they are respectively the coldest and the hottest months when the academic activity is intense.

4.1   System with the Air Conditioner Plant Managed by the Proposed Power Management System

This section shows the results obtained in the same conditions described in the previous sections but with the presence of an air conditioner plant managed by the proposed auction based power management system. This system automatically shut down the air conditioner in the rooms where there are no people.

Figure 2 and Figure 3 show the obtained results respectively in January and July. In both figures there is a green line highlighting the periods of time when the analysed room (lecturer theatre) wins an auction (and so, it can turn on its air conditioner).

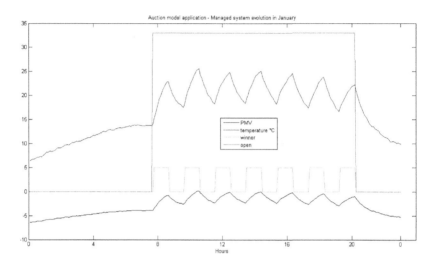

Fig. 2 Daily system evolution in January using an air conditioner plant managed by the auction based power management system

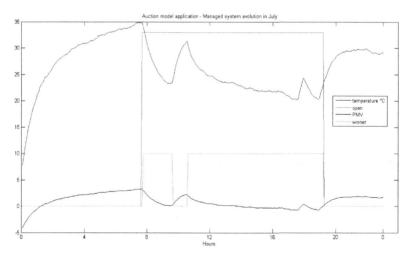

Fig. 3 Daily system evolution in July using an air conditioner plant managed by the auction based power management system

In order to show a quantitative evaluation of the effectiveness of the proposed system, Table 2 reports the mean values of temperature, the PMV index obtained in each room. In January it is possible to save the 33% of energy obtaining a good mean level of wellness in all the rooms. Only in library there is a mean value of PMV equal to -1.12 that is out of the optimal range ([-0.5; +0.5]), this value corresponds to a percentage of satisfied people greater than 75%. In July, in order to obtain good value of wellness it is necessary to increase the quantity of used power energy. Indeed, the percentage of saved energy is 25%.

Table 2 Mean values of temperature and PMV obtained in each room and percentage of saved power in January and July

January	PMV	Temp. (°C)	July	PMV	Temp. (°C)
Bar	-0,53	23,58	Bar	0,29	23,98
Library	-1,12	21,86	Library	0,99	26,57
Classroom 1	0,66	27,05	Classroom 1	-0,08	22,64
Auditorium	-0,28	24,30	Auditorium	0,04	23,08
Classroom 2	0,27	25,92	Classroom 2	0,25	23,84
Classroom 3	0,01	25,16	Classroom 3	-0,02	22,86
Classroom 4	0,36	26,17	Classroom 4	-0,06	22,72
Classroom 5	0,56	26,76	Classroom 5	-0,16	22,34
Classroom 6	0,30	25,99	Classroom 6	0,60	25,12
Classroom 7	0,47	26,50	Classroom 7	0,09	23,26
Classroom 8	0,49	26,55	Classroom 8	0,03	23,02
Classroom 9	0,73	27,23	Classroom 9	0,00	22,92

5   Discussion and Conclusions

The paper presents an innovative intelligent DSM system managed according to an auction based multi-agent system. It has been tested within a project aiming at defining a technology monitoring and management of power consumption for real civil applications. The solution adopted for the application discussed, based on the characterization of the habits of individual users, allows to obtain the best control practice.

The proposed system showed a good capability to optimize energy consumption in the test case addressed. The results shown in Table 2 demonstrate the effectiveness of the rules adopted for the proposed system in energy saving while maintaining a good level of comfort to the users. In particular, in January, it was possible to save the 33% of energy obtaining a good average level of wellness in all the rooms. The worst results are obtained in the library room where there is a mean value of PMV index equal to -1.12 that is out of the optimal range ([-0.5;

+0.5]). In July, in order to obtain a good value of wellness, it is necessary to increase the quantity of used power energy. Indeed, the percentage of saved energy is 25%. It should be highlighted that these two months are the most critical ones from a climatic point of view in the analysed area. For the other months, it is possible to achieve also better results in terms of energy saving.

This model may become the starting point of a path aiming at satisfying the needs of individual users in order to minimize the overall energy consumption of any requirement.

Acknowledgments. The innovative contents described in this paper are disclosed after the permission of TES Energy System S.r.l company, which committed to Laboratorio Kad3 S.c.a.r.l. the research project called "Power Consumption Monitoring and Optimization (Po.Co.Mo.Pti.)".

References

[1] Nadel, S.: Utility Demand-Site Management Experience and Potential - A Critical Review. Annual Review of Energy and the Environment 17, 507–535 (1992)

[2] Di Lecce, V., Pasquale, C., Piuri, V.: A Basic Ontology for Multi Agent System Communication in an Environmental Monitoring System. In: Proc. IEEE CIMSA, Boston, MA, pp. 45–50 (2004)

[3] Di Lecce, V., Calabrese, M., Soldo, D.: Semantic Lexicon-based Multi-Agent System for Web Resources Markup. In: Proc. ICIW 2009, Mestre, Italy, pp. 143–148 (2009)

[4] Di Lecce, V., Calabrese, M.: Describing non-selective gas sensors behaviour via logical rules. In: Proc. IEEE/ACM SENSORCOMM 2011, Nice/Saint Laurent du Var, France, pp. 6–11 (2011)

[5] Koppensteiner, G., Merdan, M., Lepuschitz, W., Reinprecht, C., Riemer, R., Strobl, S.: A Decision Support Algorithm for Ontology-based Negotiation Agent within Virtual Enterprises. In: Proc. IEEE FITME 2009, Sanya, China, pp. 546–551 (2009)

A Dynamic Agent-Based Modeling Framework for Digital Business Models: Applications to Facebook and a Popular Portuguese Online Classifieds Website

Aneesh Zutshi, António Grilo, and Ricardo Jardim-Gonçalves

Abstract. This paper presents an Agent Based Modeling Framework that can be used to model any internet based business. The model captures the unique characteristics that define how online users interact, share information, and take product adoption decisions. This model can be used to simulate business performance, make business forecasts, and test business strategies. To demonstrate the model, we have applied it to Facebook as well as a leading Portuguese online classifieds website – CustoJusto.pt. Through a number of cases, we have simulated the growth forecasts, its impact by changes in pricing, and changes in the Business Model itself.

Keywords: Digital Business Models, Agent Based Modeling, Business Forecasting, Business Simulation, Facebook, CustoJusto.

1 Introduction

Digital Businesses are the most dynamic of industry segments, and have expanded dramatically over the last decade. Digital businesses encompass the entire gamut of ventures, from online sale of products and services to social collaboration platforms, and have become a vital engine for the new economy.

In this paper, we first make a review of the characteristics that make a Digital Business different from traditional businesses. Through a Business Model approach we develop a generic framework that can be used to represent a Digital Business. This framework, A Dynamic Agent Based Modeling Framework

Aneesh Zutshi · António Grilo · Ricardo Jardim-Gonçalves
Universidade Nova de Lisboa
Lisbon, Portugal

P.-J. Benghozi et al. (eds.), *Digital Enterprise Design & Management*,   105
Advances in Intelligent Systems and Computing 261,
DOI: 10.1007/978-3-319-04313-5_10, © Springer International Publishing Switzerland 2014

(DYNAMOD), further incorporates Agent Based Modeling techniques to model real life business scenarios. DYNAMOD provides a framework to assist business managers develop and test their models in a variety of market scenarios. DYNAMOD will enable the simulation of current market conditions, and enable the visualization of market adoption and growth of a particular business offering, thus enabling the development of a sound Business Model with a much greater chance of success. Finally, we demonstrate some of the capabilities of this model through its application to Facebook and a popular portuguese online classifieds website- CustoJusto.pt.

2 Digital Business Characteristics

Digital businesses refer to the provisioning of products or services through the Internet. They have been the most dynamic entrepreneurial area, requiring agile and innovative business models. Digital businesses have flourished and failed at a scale and pace never seen before by any other traditional business sector. Despite numerous failures, the phenomenal success of new ventures like Google and Facebook have shown that disruptive business models have the potential to exponentially gain market share and expand virally. Here we discuss some of the key characteristics that form the defining features of digital businesses.

2.1 Free and Freemium Models due to Low Cost Per User

While costs of infrastructure and servicing of clients has increases over time for traditional businesses, these costs have been falling rapidly in the case of digital businesses. The key infrastructure costs – Storage, Processing and Bandwidth, have continued to follow Moore's Law and have steadily plummeted. As companies have embraced economies of scale, the servicing cost per client have been tending to zero (Teece, 2010). This has led to the growth of Free and Freemium business models, where the end user receives tangible products or services without a cost.

2.2 Viral Marketing

Viral Marketing refers to the fast and infectious spread of a product across the market. It involves strategies for rapid uptake of electronic peer to peer referrals (De Bruyn & Lilien, 2008). It also involves an effective utilization of word of mouth or established networks of clients, through an excellent value proposition that is low cost or free. In electronic business, automated viral marketing strategies are sometimes used to induce viral marketing with little intervention from the users, or sometimes even without the knowledge of users. A commonly cited example is Hotmail, which became very popular due to automatically adding its web-link at the end of each email sent by a Hotmail user (Montgomery, 2001).

2.3 *Intense Competition*

Another feature of digital business is intense competition. Since entry barriers to markets tend to be low and national boundaries are not significant hurdles, the entry of new competitors takes place at a phenomenal speed. In traditional businesses, a large number of competitors would have meant a highly fragmented market share. A unique feature of digital business is that often a dominant player assumes an extremely large proportion of the market share. This is because of the low cost structure (often zero or near zero) of the digital business, where pricing cannot be used as a differentiating parameter. The network effects associated with well connected users mean that a well designed product, usually the first mover, would virally expand due to positive customer feedback (Gallagher & West, 2009).

2.4 *Difficulty in Implementing Legal Protections*

It has become extremely difficult to enforce copyright protection in digital businesses. Rampant piracy of digital products is a fine example. New business models that target at alternative sources of revenue need to be adopted (Teece, 2010). The music industry has evolved, and new models of song sales such as iTunes are being used. The software industry has been addressing this challenge by providing Software as a Service, pay as you go, and other pricing models, reducing the initial cost of purchase. Subscription based services also provide support and additional benefits, thus discouraging users to go for pirated softwares.

2.5 *Product and Marketing Integration*

Digital markets often do not provide room for drawing board development of business strategies that can be slowly tested in the market and tweaked as time goes on. For effective marketing in the digital world, marketing strategies should not be separate from product, but rather marketing should be built into the product itself (Grewal et al., 2010). Market opportunities must be quickly exploited, before a competitor catches up. This integrated approach to the digital business strategy, which incorporates product development, marketing and pricing forms the underlying principle of a digital business model.

2.6 *Network Effects*

Network effects exist when consumers derive utility from a product based on the number of other users. These effects are especially relevant for several online businesses, especially since various online products and services exhibit some form of network effects, such as social networking sites and online marketplaces. (Goldenberg, Libai, & Muller, 2010) predict a *chilling effect of network externalities*. They propose that a product with a network externality has a slower initial adoption compared to a product that does not have any network externalities. The higher growth rate due to the network effects occurs only after the product has crossed a certain adoption critical threshold.

2.7 Pricing of Information Goods/Services

The information industries have always raised challenging business model issues because information is often difficult to price, and consumers have many ways to obtain certain types of products without paying. Figuring out how to earn revenues (i.e. capture value) from the provision of information to users/customers is a key (but not the only) element of business model design in the information sector (Teece, 2010). Pricing of Digital Products is another important area of research, especially the impact of various dynamic pricing strategies and the immensely successful Freemium-based business models. Pricing of digital products often involves splitting the product into different sub-categories and re-bundling them. The unit of charge for digital products must change. Smaller units of charge, focusing on pay per use or per month subscription charges have met with success in the digital world (Docters, Tilstone, Bednarczyk, & Gieskes, 2011).

3   Agent Based Modeling Approaches to Simulate Business Environments

Traditional management guidelines are ill-equipped to help understand the digital business models and predict success and failure. New tools and techniques are necessary to help model the complex nature of online products and services. Hence we have developed a generic framework for a customizable simulation environment that can capture the dynamics of an online market, and provide Business Managers with tools to simulate and forecast, thus acting as a tool to perfect their Business Model. Online markets can be represented as a network of interconnected online users which share positive and negative feedbacks and respond to different online products and services. If the behavior of individual agents can be sufficiently well modeled, then a natural candidate for representation is Multi-Agent Based Modeling Techniques.

Agent Based is build on proven, very successful techniques such as discrete event simulation and object oriented programming (North & Macal, 2007). Discrete-event simulation provides a mechanism for coordinating the interactions of individual components or "agents" within a simulation. Object-oriented programming provides well-tested frameworks for organizing agents based on their behaviours. Simulation enables converting detailed process experience into knowledge about complete systems.

The literature review reveals that applications of Agent Based Modelling have been made to model specific areas of Business. These include prediction of financial distress (Cao & Chen, 2012), product adoption (S. Kim et al. 2011), consumer behaviour (Vanhaverbeke & Macharis, 2011), , market share (Kuhn, Courtney, Morris, & Tatara, 2010), and demand forecasting (Ikeda, Kubo, & Kobayashi, 2004). (Kuhn et al., 2010) demonstrated the possibilities of predicting market share based on certain BM attributes of Frontier Airlines. (Bellman et al., 2013) addresses the issue of capturing internet behaviour to deliver relevant advertisements. ABM approaches can also be used for modeling user response to different sources of advertising. It can also be used response modelling to identify the most critical target groups, complementing traditional approaches for the same (Lee, Shin, Hwang, Cho, & MacLachlan, 2010). Tesfatsion introduced Agent-Based Computational Economics as the computational study of dynamic economic systems modeled as virtual worlds of interacting agents. (Aliprantis, Tesfatsion, & Zhao, 2010) have applied ACE to retail and wholesale energy tradings in the Power Markets.

In this paper we extend the concept of Agent-Based Computational Economics, to develop DYNAMOD- An Agent Based Modeling Framework for online Digital Business Models. It is built upon the following areas of research:

3.1   Diffusion of Innovations

Diffusion of Innovations has been an active research area and reflects adoption decisions made by individual consumers. These decisions are made in a complex, adaptive system and result from the interactions among an individual's personal characteristics, perceived characteristics of the innovation, and social influence. Diffusion of innovation has traditionally been approached with mathematical modeling, however as computational powers increased, relatively recent attempts have been made to complement these classic approaches with Agent Based Modeling tools (Delre, Jager, Bijmolt, & Janssen, 2007; Diao, Zhu, & Gao, 2011; Stonedahl, Rand, & Wilensky, 2008). We have used these works as the basis for developing the DYNAMOD model with the application of specific characteristics to differentiate online businesses.

3.2   Word of Mouth

Word of mouth communication is more effective when the transmitter and recipient of information share a relationship based on homophily (tendency to associate with similar persons), trust and credibility. (Brown, Broderick, & Lee, 2007) conducted research on online word of mouth and report that online homophily is almost entirely independent of interpersonal factors, such as an evaluation of individual age and socio- economic class, traditionally associated with homophily. The idea of individual-to-individual social ties is less important in an online environment than in an offline one. Individuals tend to use websites as proxies for individuals. Thus, tie strength was developed between an information seeker and an information source as offline theory suggests, but the information "source" is a Web site, not an individual.

4   The DYNAMOD Framework

DYNAMOD is based on Agent Based Modeling, which enables dynamic representation of the online marketplace. Every online user that could be a potential customer for a product or service is represented as an Agent in DYNAMOD (See   ). These agents interact with each other and share information about new products and services. At the same time, they are influenced by Advertising and Social sites. The model captures these influences, and simulates their impacts in order to predict future scenarios.

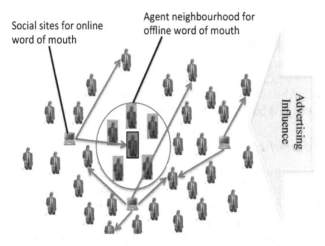

Fig. 1 Conceptual Representation of The DYNAMOD Framework

The model is customizable and extendible to implement a diverse set of Business Model components, and to make a variety of simulations.   shows a conceptual relationship of the various components the DYNAMOD Framework. The model core consists of many interacting agents that represent a market. The model includes standard variables and logics for implementing influence and satisfaction scores for each agent.

Competitor   Analysis involves introduction of competitors who can have competing influences on consumers, and then monitoring the switching behavior of consumers. **Pricing Analysis** involves introduction of various charging units, and their impacts on consumer adoption. It also involves the introduction of

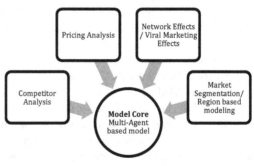

Fig. 2 DYNAMOD Components Architecture

Freemium Business Models into the model, and simulates the adoption of Free and Paid components of the Businesses. This module is not needed in case of Free Business Models. Businesses that have an inherent **Network Effect** or are based on **Viral Marketing** need to add additional logics that change the rate of product adoption. The **Market Based Segmentation** or **Region Based Modeling** changes the dispersion of agents in the model space, to represent different clusters of agents. This can represent different classes of customers with varying purchasing powers, or can represent customers on different continents.

5   Case Study 1 – CustoJusto.pt

CustoJusto.pt is the Portuguese subsidiary of Schibsted Classified Media (SCM), which provides services in the area of online classifieds in Switzerland, Spain, Frence Belgium, Italy and Austria. It also owns sites in Malaysia, Argentia, Mexico, Brazil and Columbia. However the focus of our study is just the Portuguese operations. . Every type of product – old or new could be advertised on the site, including but not limited to Real Estate, Rentals, Cars, Tools, etc.

In May 2013, which was the last month for which records were obtained, CustoJusto had an average of 154.623 daily unique visitors and 4750 daily unique sellers. In the same month, the number of estimated unique monthly visitors were 927.741.

The site follows a freemium model where, advertisements can be posted for free with a validity of 2 months by any registered user. A freemium revenue model has been used where most users can post for free, but at the same time it charges for some premium features. One of the services that are charged is "Editing an Ad" which costs 2 Euros per modification made to the advertisement, which we shall further explore in Section 5.2.2. Also, the company is currently spending 80% of its resources towards marketing costs which includes Television and Online Advertisements because of an advertising blitzkrieg launched by its closest rival.

5.1   Modeling Process

The DYNAMOD model is generic and can be modeled using any modeling language. However we have used NETLOGO 5.0.3 as the development environment to model this case study. This provides us with an extensive library to program the agent behavior, environmental constraints and the modeling parameters. It also provides us with a graphical interface to review the simulation results.

Detailed interviews were conducted with the Business Managers to understand the nature of the Business. As per http://www.internetworldstats.com, Portugal has

5950 Thousand users. Also, according to http://www.pordata.pt, average household size in Portugal is 2,6. In our assessment, buying and selling of household items, is typically generally done by one member for each household. Hence the typical market size for such a free classified service would typically be 5950 thousand / 2,6 = 2288 thousand users. Using Agent Compression techniques to run the model with a smaller number of Agents (Wendel & Dibble, 2007), we approximate 1 agent for every 1 thousand users. Hence the simulation was initialized to 2288 agents.

To initialize the agents, we conducted an online sample survey in June 2013 to obtain a random sample of user opinion towards the website and their response to different sources of advertising. The questionnaire was sent to 2000 users and we received a response from 157 users. Mean and standard deviations were calculated for every response.

5.2   Modeled Scenarios

The model has a very high offline Word of Mouth factor as was apparent from our sample surveys. It also benefits from a high network effect coefficient since having a large number of Sellers attract a large number of Buyers and vice versa. Hence we assign the network effect coefficient (C_{NE} as +0.5).

We obtained data for number of users accessing the site every month to estimate the monthly users to the site for 53 months from Jan 2009 to May 2013. We divide the data into two sets: From Jan 2009 to July 2011 as the initialization set and from August 2011 to May 2013 as the validation set. The first set of data is used to initialize the model. In this phase, the data is used to adjust the key model coefficients, such as Influence Coefficient ($A_i^{Influence-Coefficient}$, which determines the rate of change of influence for a non-customers in a population) and the Radius of Influence (Radius around each agent where neighborhood influence can propagate). These coefficients are determined to ensure that the model prediction is closest to the model initialization data. Subsequently, the model is allowed to run till July 2015. The data from August 2011 till May 2013 is used to validate the results of the Model. This validation is shown in scenario one.

5.2.1   Scenario 1 – Simple Forecast

The real data as well as the model prediction is plotted in Figure 3. The Data is highly seasonal. This is because it was not possible to obtain the actual number of users registered, because only sellers need to register on the site while buyers need not register on the site. Hence we have used unique monthly visitors to the site to estimate the value of actual number of customers.

Fig. 3 DYNAMOD Model Run Results for CustoJusto.pt

Mean Absolute Percent Error during Initialization phase

$$= \frac{(\Sigma_{Jan\,09}^{Jul\,11}(|Ct-CAt|/CAt)\%)}{Count-Months\,(Jan\,09-Jul\,11)} = 10\%$$

Mean Absolute Percent Error during Validation phase

$$= \frac{(\Sigma_{Aug\,11}^{May\,13}(|Ct-CAt|/CAt)\%)}{Count-Months\,(Aug\,11-May\,13)} = 11\%$$

While the main aim of the Model is not to be used as a pure forecasting tool, a good forecast ensures that the model is efficiently configured and ready for making further analysis. To validate the forecasting efficiency of the current model we compare it with ARIMA Simple Exponential Smoothing forecasting. Figure 3 shows the plot of ARIMA with simple exponential smoothing alongwith DYNAMOD forecast and the real historical values.

Mean Absolute Percent Error during Validation phase for ARIMA = 11%

This demonstrates that the model error is same as the ARIMA simple Exponential Smoothing predictions, and the Agent Based Model can be considered as well initialized and having a good forecast.

5.2.2   Scenario 2 - Analysing Price Sensitivities (Editing an Ad)

"Editing an Ad" is a premium service that is charged at €2 per edit. However it is not a popular service because instead of editing old ads, users can always post new ads for free. Once the model is configured it is possible to change the value of this service to view the customer adoption rate.

Calculating Maximum Profitability

We shall program the model to modify "Editing the add" cost and note the number of users of the service after a 6 month period. We shall conduct the model run for 6 sets of values: €0,50, €1, €1,5, €2, €2,5, and €3. As indicated by our survey we will use a typical usage rate of once in 2.3 months.

As can be seen from the above analysis, Revenue Maximization can be made by changing the price of "Editing an Ad" from €2 to €1. This insight can also be sought from direct mathematical analysis of the current data. However, that would only provide us with a snapshot of the current price sensitivity. By incorporating the results into a Dynamic model, we can visualize the long term impacts of a change in the pricing of a particular service.

Table 1 Price Sensitivity Simulation for Editing an Ad

	€0,5	€1	€1,5	€2	€2,5	€3
No. of Users	165981	156864	95672	75961	33238	13431
Revenue	€36082	€68201	€62394	€66053	€28902	€11679

Similar calculations can also be made for other premium services, but have not been shown since the objective of this paper is an introduction to the capabilities of the DYNAMOD model.

5.2.3   Scenario-Modifying Advertising Expenditure

The high Value of $\bar{A}_i^{\text{Influence-TV}} = 5.96$ and $\bar{A}_i^{\text{Influence-online-ads}} = 5.51$ shows that users are highly influenced by online ads. Hence it will be interesting to see the impact of change in Advertising Budget which is currently 80% of the total costs. Suppose we reduce this expenditure share to 40% of total expenses, thus bringing down overall company expenses.

$\text{Cost}^{\text{Ad-TV}} = 0.4$ from 0.8, and, $\text{Cost}^{\text{Ad-Online}} = 0.4$ from 0.8

We make the change from on running the model from June 2013, and observe the value of number of users 6 months later, in Dec 2013. According to the model,

Count $\sum_{Dec\,13} Ai$ where $(\text{Cost}^{\text{Ad-TV}}$ & $\text{Cost}^{\text{Ad-Online}} = 0.8) = 1194$

Count $\sum_{Dec\,13} Ai$ where $(\text{Cost}^{\text{Ad-TV}}$ & $\text{Cost}^{\text{Ad-Online}} = 0.4) = 1189$

Surprisingly, the difference in number of users 6 months after the 50% reduction in advertising expenses is not much according to the model. While, its true that Advertising plays a major role, Word Of Mouth, especially offline word of mouth has a far greater influence, $\bar{A}_i^{\text{OfflineWOM-Friends}} = 6.93$. Also the model has already reached 50% market penetration, hence the critical mass has already been achieved, and the site can sustain itself due to strong Network Effects and an effective Word of Mouth influence. According to the model predictions, it could be wise to spend lesser on Advertising and use those resources in the development of new service

6   Case Study 2 – Facebook

Facebook is an extremely popular social networking site that enables users to share messages and posts. It exhibits a strong Network Effect, because the more friends a user has on Facebook, the more useful Facebook can be in communicating with them. For the purpose of this simulation, we have used data from Stock Market Declaration of the number of active clients from June 2009 to September 2012. Data from June 2009 till June 2011 was used to initialize the Agent Based Model. Records from September 2011 to September 2012 were used as validation data. A total of 112 people participated in a sample survey out of which 102 were Facebook users. While the survey was conducted online, most respondents were from Lisbon, Portugal. A potential global user base of 1400 million users were estimated.

6.1 Scenario 1 – Simple Forecast

To test the accuracy of forecast by the DYNAMOD model, we also used another forecasting tool. Since the data untill June 2011 were mostly linear, with no cyclical trends, we decided to also use linear regression for forecasting future adoption rates.

The Root Mean Square Error for the initialization data on the linear regression line was 1.36 indicating a very close model fit, while the DYNAMOD line had a Root Mean Square Error of 2.92. Despite the closer fit by the linear regression line, the results from the forecasts show an interesting reversal of trends. The actual growth of Facebook active users slows down over the validation period. Hence the regression line moves away from the actual growth line. However the DYNAMOD model is able to predict the slowdown and closely follows the actual growth line. For the Validation Period between Sep 2011 and Dec 2012, the Regression line prediction has a root mean square error of 13.7 while the DYNAMOD model has a RMS error of 2.92. This demonstrates that DYNAMOD can be leveraged for business growth simulations, in this case, a very mature business which has already reached a majority of global users.

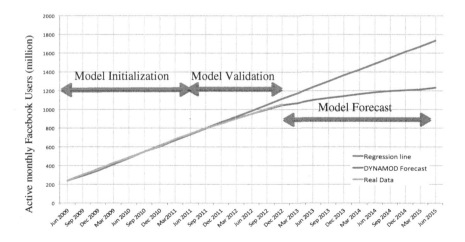

Fig. 4 DYNAMOD Forecasts for Facebook

7   Conclusions

Only a limited number of models and scenarios have been described in this paper due to space limitations. They were chosen to provide an introductory demonstration to the potentials of DYNAMOD for Business Analysis and Forecasting. It is not merely a forecasting tool, but a comprehensive instrument for analyzing and representing the business scenarios and market conditions.

This paper proposes an Agent Based Modeling Framework that can be applied to a variety of online Businesses to analyze, evaluate and predict business growth. The Framework captures the way that online users are organized and communicate with each other to share "word of mouth" and in the process propagate the adoption of a product or a service. The model also captures other novel elements of digital businesses such as network effects, various pricing models including Freemium, and various marketing strategies such as viral marketing. The Framework can be applied for Business to Business(B2B), Business to Consumer(B2C), and Consumer to Consumer(C2C), all types of business scenarios. This framework can be applied to businesses in early stages, middle stages and highly mature stages.

Agent Based Model applications in the business world, especially Digital Businesses is still in its infancy. At the same time, digital businesses, which are now steering the growth in economies, are still not academically explored. By merging these two fields of research, we aim to provide an academically validated tool to help online Entrepreneurs as well as digital business managers, better manage their online businesses.

References

Aliprantis, D., Tesfatsion, L., Zhao, H.: An Agent-Based Test Bed for the Integrated Study of Retail and Wholesale Power System Operations....technology for energy systems (2010)

Bellman, S., Murphy, J., Treleaven-Hassard, S., O'Farrell, J., Qiu, L., Varan, D.: Using Internet Behavior to Deliver Relevant Television Commercials. Journal of Interactive Marketing 27(2), 130–140 (2013), doi:10.1016/j.intmar.2012.12.001

Brown, J., Broderick, A.J., Lee, N.: Word of mouth communication within online communities: Conceptualizing the online social network. Journal of Interactive Marketing 21(3), 2–20 (2007), doi:10.1002/dir.20082

Cao, Y., Chen, X.: An agent-based simulation model of enterprises financial distress for the enterprise of different life cycle stage. Simulation Modelling Practice and Theory 20(1), 70–88 (2012), doi:10.1016/j.simpat.2011.08.008

De Bruyn, A., Lilien, G.L.: A multi-stage model of word-of-mouth influence through viral marketing. International Journal of Research in Marketing 25(3), 151–163 (2008), doi:10.1016/j.ijresmar.2008.03.004

Delre, S.A., Jager, W., Bijmolt, T.H.A., Janssen, M.A.: Targeting and timing promotional activities: An agent-based model for the takeoff of new products. Journal of Business Research 60(8), 826–835 (2007), doi:10.1016/j.jbusres.2007.02.002

Diao, J., Zhu, K., Gao, Y.: Agent-based Simulation of Durables Dynamic Pricing. Systems Engineering Procedia 2, 205–212 (2011), doi:10.1016/j.sepro.2011.10.024

Docters, R., Tilstone, L., Bednarczyk, S., Gieskes, M.: Pricing in the digital world. Journal of Business Strategy 32(4), 4–11 (2011), doi:10.1108/02756661111150927

Gallagher, S., West, J.: Reconceptualizing and expanding the positive feedback network effects model: A case study. Journal of Engineering and Technology Management 26(3), 131–147 (2009), doi:10.1016/j.jengtecman.2009.06.007

Goldenberg, J., Libai, B., Muller, E.: The chilling effects of network externalities. International Journal of Research in Marketing 27(1), 4–15 (2010), doi:10.1016/j.ijresmar.2009.06.006

Grewal, D., Janakiraman, R., Kalyanam, K., Kannan, P.K., Ratchford, B., Song, R., Tolerico, S.: Strategic Online and Offline Retail Pricing: A Review and Research Agenda. Journal of Interactive Marketing 24(2), 138–154 (2010), doi:10.1016/j.intmar.2010.02.007

Ikeda, Y., Kubo, O., Kobayashi, Y.: Forecast of business performance using an agent-based model and its application to a decision tree Monte Carlo business valuation. Physica A: Statistical Mechanics and its... 344(1-2), 87–94 (2004), doi:10.1016/j.physa.2004.06.093

Kim, S., Lee, K., Cho, J.K., Kim, C.O.: Agent-based diffusion model for an automobile market with fuzzy TOPSIS-based product adoption process. Expert Systems with Applications 38(6), 7270–7276 (2011), doi:10.1016/j.eswa.2010.12.024

Kuhn, J.R., Courtney, J.F., Morris, B., Tatara, E.R.: Agent-based analysis and simulation of the consumer airline market share for Frontier Airlines. Knowledge-Based Systems 23(8), 875–882 (2010), doi:10.1016/j.knosys.2010.06.002

Lee, H., Shin, H., Hwang, S., Cho, S., MacLachlan, D.: Semi-Supervised Response Modeling. Journal of Interactive Marketing 24(1), 42–54 (2010), doi:10.1016/j.intmar.2009.10.004

Montgomery, A.: Applying quantitative marketing techniques to the Internet. Interfaces (2001)

North, M., Macal, C.: Managing business complexity. Discovering strategic solutions with agent-based...(2007)

Stonedahl, F., Rand, W., Wilensky, U.: Multi-Agent Learning with a Distributed Genetic Algorithm Exploring Innovation Diffusion on Networks. In: Proceedings of the ALAMAS+ALAG Workshop at AAMAS 2008, pp. 2–9 (2008)

Teece, D.J.: Business Models, Business Strategy and Innovation. Long Range Planning 43(2-3), 172–194 (2010), doi:10.1016/j.lrp.2009.07.003

Vanhaverbeke, L., Macharis, C.: An agent-based model of consumer mobility in a retail environment. Procedia - Social and Behavioral Sciences 20, 186–196 (2011), doi:10.1016/j.sbspro.2011.08.024

Wendel, S., Dibble, C.: Dynamic agent compression. Journal of Artificial Societies and Social...(2007)

The Impact of 3D Printing Technologies on Business Model Innovation

Thierry Rayna and Ludmila Striukova

Abstract. There is a growing consensus that 3D printing technologies will be the next major technological revolution. While a lot of work has already been carried out as to what these technologies will bring in terms of product and process innovation, little has been done on their impact on business model innovation. Yet, history has shown that technological evolution without adequate business model innovation is a pitfall for many businesses. The contribution of this article is threefold. First, it combines the existing literature on business model innovation in an integrated 'inside-outside' framework of business model innovation. Secondly, the changes brought about by 3D printing technologies to the business model components are investigated. Finally, this article shows that in addition to enabling business model innovation, 3D printing technologies have the potential to change the way business model innovation is done, by enabling adaptive business models and by bringing the 'rapid prototyping' paradigm to business model innovation itself.

1   Introduction

While everyone agrees on the critical importance of innovation, businesses often fail to innovate successfully because of a too narrow view of innovation. Indeed, when thinking about innovation, technological innovation (which results in product or process innovation) is often the first thing that comes to mind. Sadly, it is also far too often the *only* thing that comes to mind. Yet, technological innovation can only be valuable with an adequate business model. Countless firms with technological leadership have failed because of an unsuitable business model, while less

Thierry Rayna
ESG Management School, 25 rue Saint Ambroise 75011 Paris, France
e-mail: trayna@esg.fr

Ludmila Striukova
University College London, Gower Street London WC1E 6BT, UK
e-mail: l.striukova@ucl.ac.uk

P.-J. Benghozi et al. (eds.), *Digital Enterprise Design & Management,*   119
Advances in Intelligent Systems and Computing 261,
DOI: 10.1007/978-3-319-04313-5_11, © Springer International Publishing Switzerland 2014

innovative firms have achieved market dominance simply because of a better business model.

Furthermore, technological innovation might even sometimes endanger a successful firm (e.g. digitisation and the major record labels). Indeed, technological innovation often requires to completely rethink a firm's business model. Instead, technological innovation often provides firms with a false sense of safety, which might eventually be lethal. Meanwhile, technological followers often realise that the best way to catch up with the leaders is through business model innovation. Recent history has shown that business model innovation is, indeed, a powerful tool. The victories of Apple on the 'device' markets (iPod, iPhone, iPad) and online music (iTunes Store) are a testimony to the fact that a well thought business model innovation may be far more potent than market dominance or technological or product leadership (Rayna et al., 2009; Rayna and Striukova, 2009).

Amongst the recent technological innovations, 3D printing (or 'additive manufacturing') has been deemed as a very promising one. In 2013, U.S. President Barack Obama mentioned the critical role of 3D printing in strengthening manufacturing, scientific, defence and energy sectors[1]. Beforehand, Rich Karlgraad (Forbes) conjectured that 3D printing would become the "transformative technology of the 2015–2025 period" (Karlgraad, 2011). Likewise, Chris Anderson (Wired) forecasted that the "desktop manufacturing revolution [. . .] [would] change the world as much as the personal computer did" (Anderson, 2012).

Just like digitisation of other products (music, movies, books), 3D printing is going to be very disruptive, as it enables digitisation of objects. Just like what happened with other 'digitalised' industries, 3D printing is going to threaten the position of established firms and create opportunities for newcomers. In this context, business model innovation is going to play a critical role in the success or survival of firms affected by this new set of radical technologies.

This research investigates the role of 3D printing technologies in regard to business model innovation. In particular, emphasis is put on the business model components likely to be most affected by these new technologies. One of the objectives is to demonstrate that 3D printing technologies are not only enabling business model innovation, but also have the potential to considerably change the way business innovation is understood and carried out.

The first section of this article provides a classification of business model innovation and provides a novel framework which reconciles the 'inside' and 'outside' view of business innovation in the literature. The fourth section presents briefly 3D printing technologies and the main services that are currently available to businesses. Finally, the fifth section analyses the impact of 3D printing technologies on business model innovation.

---

[1] http://www.whitehouse.gov/state-of-the-union-2013

2 Categorising Business Model Innovation

Categorising business model innovation requires a thorough understanding of what business models are. The literature on business models is very abundant, with 110,000 academic works written on the subject between 2001 and 2013 and 12,500 new ones produced so far in 2013 alone[2]. Although there are differences among scholars, in particular between American and European (De Reuver et al., 2013), there is a broad consensus around four critical components of a business model: value proposition (Voelpel et al., 2004; Casadesus-Masanell and Ricart, 2010; Chesbrough, 2010; Teece, 2010), value creation (Zott and Amit, 2002; Voelpel et al., 2004; Chesbrough, 2007), value capture (Chesbrough, 2007; Holm et al., 2013), value delivery (Osterwalder et al., 2005; Abdelkafi et al., 2013; Holm et al., 2013). A fifth component, value communication, is also often thought as a critical aspect of a business model (Abdelkafi et al., 2013). The components as well as their sub-components identified in the literature are summarised in Figure 1.

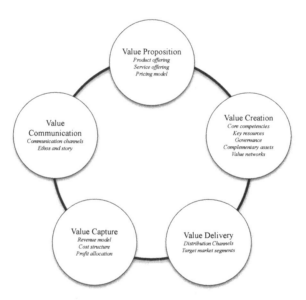

Fig. 1 Key components of business models

The most straightforward way to envisage business model innovation is to consider changes in any of these main components or sub-components (Johnson et al., 2008; Abdelkafi et al., 2013). In particular, Zott and Amit (2002), Giesen et al. (2007) and Koen et al. (2011) emphasise the role of value networks (which include all the firm partners, including customers).

---

[2] Data provided by Google Scholar.

One of the most common ways to categorise innovation is to distinguish between incremental and radical innovation (Banbury and Mitchell, 1995). The same distinction can be made for business model innovation. For Brink and Holmén (2009), radical business model innovation arrises when the business model has changed "simultaneously within more than one aspect or dimension". Likewise, Abdelkafi et al. (2013) note that modifying more than one value component at a time can lead to more radical innovations.

Besides the number of components affected by the changes, the extent of the changes also has to be taken into consideration. Indeed, according to Ho et al. (2011), the difference between incremental and radical business model innovation relates both to the number of business model components affected, but also to the degree of innovation. When both are high, business model innovation is radical. When both are low, it is incremental. Brink and Holmén (2009) also note that radical innovation necessarily leads to many simultaneous changes in the business model. Likewise, Voelpel et al. (2004) mentions that radical business innovation is highly disruptive for the firm itself and its key components (core structure, governance, etc.).

The problem of this classification is that there is a large 'grey' area when one of these two criteria is high and the other is low (e.g. high degree of innovation affecting a few components of the model, low degree of innovation affecting many components). For this reason and in opposition to this 'inside view' of business model innovation (based on components), other authors consider, instead, the external aspects of business model innovation. In this case the radicalness of business model innovation is assessed based on its effect on clients, markets and industry.

Johnson et al. (2008) mention *de-novo* business models, which are not only new for the company, but also "game-changing for the industry or market". Likewise, Zott and Amit (2002) define radical business model innovation as a novel business model that leads to the creation of new market (e.g. eBay). However, radical business model innovation does not necessarily 'automatically' create new markets, but, instead, creating new markets may be needed because radical business model innovations are sometimes simply too radical for their own market (Treacy, 2004).

Creating new markets is not a necessary condition for business model innovation to be disruptive. Changes in existing markets is also a consequence of radical business model innovation. To this respect, Giesen et al. (2007) consider both redefinition of the industry in which the firm operates and horizontal move to new industries as critical aspects of business model innovation. Likewise, Koen et al. (2011) categorise business model innovation according to changes in the value network. Incremental business model innovation tend to keep the same customer base, while more innovative changes enable to capture existing customers which are not yet customers of the firm (clients of competitors). Finally, the most radical business model innovations enable to attract non-customers, hereby creating new markets.

When combining these 'internal' and 'external' views of business model innovation, it is important to keep in mind the difference between radical innovation and disruptive change. Indeed while market/industry disruption is generally associated with radical innovation, this is not necessarily always the case. Indeed, incremental innovation can lead to radical change, just as radical innovation can reveal itself as insignificantly disruptive (Rayna and Striukova, 2009). The same is also true for business model innovation. For instance, when moving horizontally to existing markets, a firm may become highly disruptive for the firms on that market, although the core of its business model will not really change. Likewise, radical business model innovation may only affect the very same consumer base as before.

Fig. 2 Inside-outside view of business model innovation

Figure 2 integrates these two different views of business model innovation. The dotted arrows symbolise the loose relationship between radical innovation and disruption and the fact that business model innovation, whether incremental or radical, may lead to a wide range of outcomes on the market, some very disruptive, others not. Furthermore, profitability resulting from business model innovation has to be taken into account. Indeed, as noted by (Amit and Zott, 2010), subtle changes to business models might not be disruptive, but, nonetheless, be profitable.

3   A Introduction to 3D Printing Technologies

3D printing is a form of 'additive' manufacturing, where a three-dimensional object is 'printed' (built) by adding layer after layer of a particular material. This differs from the more usual 'subtractive' (when an object is cut out from the raw material) or moulding/die-casting (when liquefied material is placed into a mould) forms of manufacturing. The first stage of 3D printing involves creating a digital model of the object to be printed. This is usually done with generic 3D modelling software (some of which are available for free) or using dedicated software provided by 3D printing services (e.g. Thingiverse, Shapeways or Sculpteo). 3D scanners can also be used to automatically create a model of an existing object (just like 2D scanners are used to digitise photos, drawings or documents). When an object is printed, the 3D model of the object is discomposed into successive layers that are printed one at time.

The most frequently used material for 3D printing is plastic, but wood, metal alloy, salt, ceramics and even sugar and chocolate can be used to print. Currently, most printers can only print with one material at a time, but it is only a matter of time before several materials can be used simultaneously. The Objet500 Connex (sold at $250,000) can already print from more than 100 materials (up to 14 simultaneously) and manufacture items which are at the same time both rubber and rigid, opaque and transparent. The range of objects that can be manufactured with 3D printers is very wide and is constantly growing: robots, body parts (organs), prosthetics, art, food items, musical instruments, furniture, clothes. 3D printers can be even used to print other 3D printers.

While 3D printing technologies were, originally, intended exclusively for industrial use, the constant decrease in cost has put them within reach of SMEs and individual entrepreneurs. With home 3D printers now being available for less than $1000 (the cheapest printer, the Buccaneer, costs $350), 3D printing is progressively becoming a technology any business, small or large, can afford and a number of companies have already started to integrate 3D printing into their business model.

Beyond their usage by firms, there is a growing trend of using 3D printing in consumer markets. While originally home 3D printing was often dismissed as a hobbyist activity, the entry of major players in this market tends to demonstrate otherwise. In May 2013, Staples became the first major U.S. retailer to sell 3D printers. Amazon followed the trend in June 2013, when it opened a 3D printing section, selling printers, plastic filament, books, software, parts and supplies. In July 2013, High Street consumer electronic retailer Maplin also started to sell 3D printers, consumables and accessories in its 205 stores throughout the UK.

The same month, eBay announced its new iPhone application called eBay Exact which enables users to browse and buy customisable print-on-demand merchandise from three 3D printing companies: MakerBot, Sculpteo, and Hot Pop Factory. Selfridges, the UK high-end department store will be opening, in partnership with the 3D printing service iMakr, a Christmas shop where customers can print in store, buy 3D printers and 3D scan objects. Tesco, the UK grocery chain and one of the worlds largest retailers, is also thinking about introducing 3D printing services, such as printing spare parts for items that customers had already bought while they are shopping in the store[3].

While not every business or home have (yet) their own 3D printers, a growing number of services related to 3D printing (most of them online) are offered to consumers and businesses. Companies like Ponoko (the first mover, opened in 2007),

---

[3] https://www.tescoplc.com/talkingshop/index.asp?blogid=124

Sculpteo and Shapeways operate a marketplace service where companies can sell the 3D models of their products directly to customers. The physical object can then either be printed by the marketplace for the consumer or directly by the consumer at home. If consumers do not have yet their own printer, Cubify Cloud, in addition to its marketplace and printing services, also offer to ship 3D printers directly to consumers.

In addition to these rather versatile services, there are also companies specialising in printing activities. Two of them, iMakr and Makebot, even have physical stores and are, thus, the 3D equivalent to the traditional print store. Most of these services offer users assistance with the creation of their 3D object (for instance by converting a 2D drawing into 3D). Services like MakeXYZ and AdditiveHabitat provide a market place for 3D printers, where users can locate 3D printers located next to them and get a quote from the owner of the printer for the particular object they want to print.

Finally, online platforms, such as Additer and Kraftwürx, enable crowdsourcing of both design and manufacturing. Businesses and consumer alike can use these platforms even when they only have a faint idea of what they want to manufacture (and of how to manufacture it). The elements of the 'crowd' will team up to offer designs, materials (Kraftwürx offers over 70 different materials), the result being printed at nearby location.

In addition to these services, an increasing number of consumers and businesses make the choice to lease or own their own 3D printers, a trend which has rapidly accelerated over the past months.

4 How 3D Printing Can Revolutionise Business Model Innovation

Over the past few years, it has become clear that 3D printing technologies will have a very large (and disruptive) impact on the economy. Additive manufacturing will, undoubtedly lead to significant product and service innovation. However, this should not hide the fact that these same technologies have the potential to considerably affect business model innovation.

There are, in fact, two broad ways in which 3D printing can have impact on business model innovation. The first one relates to how this new set of technologies can change the different business model components (presented in Fig. 1). The second one, probably more subtle, but potentially more radical, is that 3D printing technologies have the potential to actually change the way business model innovation is done.

4.1 Innovation in Business Model Components

In section 2, it was shown that a key aspect of business model innovation relates to changes in the different business model components. Both the extent of changes in the components and the number of components being changed can result in radical innovation.

In the case of many technological innovation, the principal business model component which is affected is *value proposition*, as technological innovation leads to product and service innovation. Although 3D printing technologies have already led to product and service innovation, their main impact is more likely to relate to the *value creation* component and, in particular its *value network* subcomponent. Indeed, one of the key aspects of 3D printing technologies is that they enable large-scale mass customisation. As a result of the co-creation process between customers and firms, the value of the resulting product is higher than for a mass-produced product. By taking an active part in the creation process, customers become a far stronger element in the Value Network and enable more value to be created.

A second element, which also relates to value network, is crowdsourcing. Crowdsourcing has already led to significant business model innovation, in some cases even to an entirely new form of business models (e.g. Kickstarter, Threadless). However, 3D printing enables to take this concept one step further. Indeed, so far crowdsourcing has been restricted to the idea/design stages of the production process. 3D printing technologies make it possible to apply the crowdsourcing paradigm to the manufacturing stage of the process. For instance, services such as Additer, Kraftwürz and MakeXYZ enable businesses to crowdsource the manufacturing of their products using various materials and finish qualities (printers available through these services range from the basic plastic home printer to industrial grade alloy printer). To this respect, the network of 3D printers available to firms can act as a valuable *complementary asset* and be integrated fully in the business model.

Another key business model component affected by 3D printing is, obviously, *value delivery*. Indeed, by enabling customers to manufacture at home (or in local print shops), 3D printing can potentially significantly alter *distribution channels*, creating new ones alongside traditional ones. For instance, accessories (e.g. smartphone cases) companies can, in addition to having their products mass-manufactured, use one of the many online 3D printing services (e.g. Cubify Cloud, i.Materialise, Ponoko, Sculpteo, Shapeways) to sell their products to consumers. Consumers can then either print the product at home (if they own a 3D printer), have it printed and delivered by the online 3D printing service, or have it printed in a local printshop (such as iMakr in the UK, MakeBot in the U.S.). Some of these online services, such as Cubify Cloud, even develop distribution channels further, as they offer consumers to purchase 3D printers and print 3D objects at home.

A further change in value delivery brought about by 3D printing relates to *target market segments*. Indeed, whereas until now niche market segments were often neglected, because of the high initial cost of manufacturing (one does not set up a production line just for a few units), 3D printing enables to serve niche markets regardless of how small they are. It enables, in a way, to monetize the 'long tail'. Indeed, set-up costs for 3D printing manufacturing are very low and it is only when a significantly high number of (presumably standardised) units needed to be produced that mass production becomes more worthy than 3D printing.

A recent example of niche market enabled by 3D printing is Square Helper, a plastic widget which prevents the Square card reader[4] from spinning when the credit card is swiped. The entrepreneur behind Square Helper has now sold over 1,000 units, which were produced using his own home 3D printer, at $8 a piece[5]. Since it was impossible to know in advance what the demand for such a widget would be, it would not be possible to mass produce the widget, as it would require a pre-commitment on quantity, which would be far too risky.

The demand for niche products has been demonstrated over the past few months by the success of Kickstarter (and similar crowd-funding platforms) projects (some raising over a million USD for electronic devices and accessories). However, the key issues of these projects is their lack of scalability: a significant number of units have to be purchased before production starts and once the initial batch has been produced it is often impossible to order more units (except if a second project is launched and there is enough demand the second time around). 3D printing technologies enable remove these two constraints and to fully exploit niches.

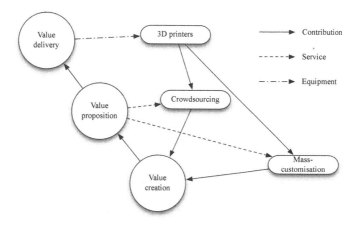

Fig. 3 Positive feedback loop between business model components

An important effect of 3D printing technologies on business model innovation is that they create a potentially positive feedback loop between value creation, value proposition and value delivery (Fig. 3). Indeed, crowdsourcing and mass-customisation enable to increase value creation, which, in its turn, enables to improve value proposition and offer services which develop further crowdsourcing and mass-customisation. Changes in value proposition lead to changes in value delivery

---

[4] Square offers a credit card payment solutions for small businesses that consists in a card reader device which can be attached to a tablet or a smartphone.

[5] http://3dprintingindustry.com/2013/03/19/
making-money-from-3d-printing-square-helper/

that can trigger a greater adoption of 3D printers (e.g. as more mass-customised products are delivered, there are more incentives for consumers to have their own 3D printer). Greater adoption of 3D printers can develop further opportunities of crowdsourcing and mass-customisation and, hence, increase value creation.

As more value can be created with 3D printing technologies, it is important to consider the question of *value capture*. The clear positive aspect of 3D printing on value capture is that it can significantly decrease the costs (*cost structure*). Indeed, as products can be manufactured on demand, transportation costs and storage costs can be decreased. Also, although the cost of manufacturing can be higher than with mass-produced techniques, the higher cost may be passed on to consumers, who will either see a benefit in a mass-customised product or will value a quicker access to the product. Furthermore, when products are home printed, the actual manufacturing cost of the product is actually borne by consumers.

However, beyond the improvement of cost structure, value capture is most likely the business model component that 3D printing will challenge the most. Indeed, while this new set of technologies will undoubtedly lead to far more value being created, this may also result in far greater difficulties to capture both new and 'old' value. Industries that had gone digital have faced the same problem, and required innovative *revenue models* to overcome it. This is certainly where business model innovation will be most critical and this may involve radical changes in *profit allocation*. Consumers taking a significant part in the production process (from design to manufacturing and distribution), are likely to be reluctant to pay as much as before, unless they perceive that a significant value (e.g. full customisation) has been added to the product. Some companies may have to completely change their revenue model and move towards more added-value products (high-tech devices cannot be printed) or derive revenue from complementary services.

4.2 Innovation in Business Model Innovation

Besides enabling business model innovation by changing business components, 3D printing technologies also have the potential to considerably change the way business innovation is carried out. The following two sections detail these critical changes.

4.2.1 Towards Adaptive and 'Mobile' Business Models

As discussed in Section 2, ability to move one's business model horizontally to existing or new markets is a key aspect of business model innovation (this corresponds to the 'outside' view of Fig. 2). However, such kind of move is often risky, because significant investments have to be made before even entering the market. 3D printing technologies make lateral moves less risky, because products can be manufactured on demand with minimal costs. Besides being used for entering existing markets, the same strategy may be used for entirely new markets.

In addition to sideways moves, 3D printing technologies can enable firm to rapidly move upstream or downstream. For instance, firms may relinquish manufacturing to customers and focus on design and service. In contrast, design firms that were dependent on intermediaries for the manufacturing of their products may decide to take manufacturing in their own hands. This also means that firms can more easily adapt the 'length' of their business model by taking on more activities (or by giving up some of them).

Thus, 3D printing technologies enable business models to become modular and adaptable. Firms can then decide, depending on the environment to adopt a narrow (focused on one particular market) or wide, long (e.g. design, manufacturing and distribution) or short (just design) business model. Furthermore, the business model becomes fully 'mobile' and can be moved up/down or sideways, as needed (Figure 4).

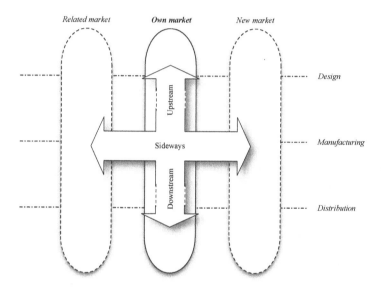

Fig. 4 3D printing enables adaptive 'mobile' business models

4.2.2 Rapid Prototyping for Business Models

Sosna et al. (2010) noted about business model innovation that firms "plan, design, test and re-test alternative business model variants until they find the one that best suits their objectives". While for businesses there is often no other choice than trial and error when it comes to business model innovation, this heuristic process generally comes at a significant cost. Many businesses do not get a second chance to experiment and firms often choose to learn from the failure of other firms rather than from their own trial and error.

In contrast, access to 3D printing technologies enable trying out various business models at a much lower cost. New ideas or design can be rapidly tested and the size of the testbed actually increases with the adoption of 3D printing technologies.

Hence, such technologies, which were used at first for rapid prototyping of objects, can also be used for rapid prototyping of business models. The ability to rapidly try and test ideas has enabled the design and manufacturing industries to significantly increase the speed of product innovation. It may well be the case that 3D printing technologies will have a similar effect on business model innovation.

5 Conclusion

Drawing on the literature devoted to business model innovation, this article has provided an integrated framework which combines both the 'inside' and 'outside' views present in the literature. This framework was then used to discuss the effect of 3D printing technologies on business model innovation. Following the 'inside' view, the potential changes to the key elements of business models were investigated. It was found that although 3D printing technologies can potentially lead to a virtuous circle of value creation, firms might find it far more challenging to capture value.

But 3D printing technologies are not only a vector of business model innovation, they can also change the way business model innovation is done, in particular this article discussed how these technologies can enable fully adaptive and 'mobile' (upstream/downstream, sideways, long or short) business models. Finally, 3D printing technologies can bring the rapid prototyping paradigm to the world of business model innovation.

3D technologies are not only disruptive to similar technologies or to technologies *per se*. They are also disruptive to the current business models, models which, in many cases, have taken a lot of time and effort to be designed. Fortunately, 3D printing technologies make it much easier to try new business models and minimised the cost for companies change markets or even their place in the value chain.

This new ability to have a very rapid rate of business model innovation creates new opportunities as well as challenges. As companies now have the ability to diversify or even change the focus of their business easily, so can competitors. Moreover, market structure is now more dynamic and key boundaries that used to exist tend to progressively being erased (e.g. consumers are becoming producers; niche market is becoming attractive to large players, not just to small ones). Chances are that the winners of tomorrow are those companies which, far from being blindsided by the new technology, will think first and foremost in terms of business model innovation.

References

Abdelkafi, N., Makhotin, S., Posselt, T.: Business model innovations for electric mobility: What can be learned from existing business model patterns? International Journal of Innovation Management 17(01) (2013)

Amit, R., Zott, C.: Business model innovation: Creating value in times of change. In: Working Paper WP-870. IESE Business School, Barcelona (2010)

Anderson, C.: Makers: The new industrial revolution. Random House Business Books (2012)

Aspara, J., Hietanen, J., Tikkanen, H.: Business model innovation vs replication: financial performance implications of strategic emphases. Journal of Strategic Marketing 18(1), 39–56 (2010)

Banbury, C., Mitchell, W.: The effect of introducing important incremental innovations on market share and business survival. Strategic Management Journal 16, 161–182 (1995)

Brink, J., Holmén, M.: Capabilities and radical changes of the business models of new bioscience firms. Creativity and Innovation Management 18(2), 109–120 (2009)

Casadesus-Masanell, R., Ricart, J.E.: From strategy to business models and onto tactics. Long Range Planning 43(2), 195–215 (2010)

Chesbrough, H.: Business model innovation: it's not just about technology anymore. Strategy & Leadership 35(6), 12–17 (2007)

Chesbrough, H.: Business model innovation: opportunities and barriers. Long Range Planning 43(2), 354–363 (2010)

Chesbrough, H., Rosenbloom, R.S.: The role of the business model in capturing value from innovation: evidence from xerox corporation's technology spin-off companies. Industrial and Corporate Change 11(3), 529–555 (2002)

Cooper, R.G.: Perspective: the innovation dilemma: how to innovate when the market is mature. Journal of Product Innovation Management 28(s1), 2–27 (2011)

De Reuver, M., Bouwman, H., Haaker, T.: Business model roadmapping: A practical approach to come from an existing to a desired business model. International Journal of Innovation Management 17(01), 1–18 (2013)

Giesen, E., Berman, S.J., Bell, R., Blitz, A.: Three ways to successfully innovate your business model. Strategy & Leadership 35(6), 27–33 (2007)

Ho, Y., Fang, H., Hsieh, M.: The relationship between business-model innovation and firm value: A dynamic perspective. World Academy of Science, Engineering and Technology 77, 656–664 (2011)

Holm, A.B., Günzel, F., Ulhøi, J.P.: Openness in innovation and business models: lessons from the newspaper industry. International Journal of Technology Management 61(3), 324–348 (2013)

Johnson, M., Clayton, C., Kagermann, H.: Reinventing your business model. Harvard Business Review 86(12), 50–59 (2008)

Karlgraad, R.: 3D printing will revive American manufacturing. Forbes (2011),
http://www.forbes.com/sites/richkarlgaard/2011/06/23/
3d-printing-will-revive-american-manufacturing/ (June 23, 2011)

Koen, P.A., Bertels, H.M., Elsum, I.R.: The three faces of business model innovation: challenges for established firms. Research-Technology Management 54(3), 52–59 (2011)

Matthyssens, P., Vandenbempt, K., Berghman, L.: Value innovation in business markets: breaking the industry recipe. Industrial Marketing Management 35(6), 751–761 (2006)

Osterwalder, A., Pigneur, Y., Tucci, C.L.: Clarifying business models: Origins, present, and future of the concept. Communications of the Association for Information Systems 16(1), 1–25 (2005)

Rayna, T., Striukova, L.: The curse of the first-mover: When incremental innovation leads to radical change. International Journal of Collaborative Enterprise 1(1), 4–21 (2009)

Rayna, T., Striukova, L., Landau, S.: Crossing the chasm or being crossed out: The case of digital audio players. International Journal of Actor-Network Theory and Technological Innovation 1(3), 36–54 (2009)

Sosna, M., Trevinyo-Rodríguez, R.N., Velamuri, S.R.: Business model innovation through trial-and-error learning: The Naturhouse case. Long Range Planning 43(2), 383–407 (2010)

Teece, D.J.: Business models, business strategy and innovation. Long Range Planning 43(2), 172–194 (2010)

Treacy, M.: Innovation as a last resort. Harvard Business Review 82(7/8), 29–30 (2004)

Voelpel, S.C., Leibold, M., Tekie, E.B.: The wheel of business model reinvention: how to re-shape your business model to leapfrog competitors. Journal of Change Management 4(3), 259–276 (2004)

Williamson, P.J.: Cost innovation: preparing for a 'value-for-money' revolution. Long Range Planning 43(2), 343–353 (2010)

Wirtz, B.W., Schilke, O., Ullrich, S.: Strategic development of business models: implications of the web 2.0 for creating value on the internet. Long Range Planning 43(2), 272–290 (2010)

Zott, C., Amit, R.: Measuring the performance implications of business model design: evidence from emerging growth public firms. Working paper 2002/13/ENT/SM, INSEAD, Fontainebleau, France (2002)

Section B
Posters

This section is dedicated to the papers that have been selected by the Program Committee to be presented as a poster during a special session of the conference DED&M 2014.

Due to the interest of the themes proposed by these works the Program Committee decided to publish their abstracts in the official proceedings.

Accelerating Innovation through Modular Design (API)

Nicolas Bry and Richard Hababou

Abstract. Open Innovation can be sustainably supported by modular components named APIs. With the cooperation assets it brings, APIs not only accelerate innovation endeavor, they generate new revenue streams, strengthen marketing campaigns, and extend your reach.

API scope is not limited to external exposure: by building organizational flexibility, it impacts collaborative work and innovation speed. Reuse capability is the killer app of API, as demonstrated in Orange and Société Générale use cases.

A successful API design is the one which helps others to take over, opening a space for creativity. Slick design is a major issue: excellence in code writing must be complemented by strategic innovation skills, and proceed on the route of 'design thinking'. Designing an API as a truly accomplished innovation component is a cultural achievement.

To conclude, let's look up for the best team organization to design sharp APIs. It's a combination of autonomy of the innovation team, and shrewd management of components portfolio, supported by an innovation framework. Amazon has plowed the land ahead: lets' get moving!

Nicolas Bry
Innovation Marketing Technology/OV
Orange, France

Richard Hababou
Innovation Director
Société Générale, France

P.-J. Benghozi et al. (eds.), *Digital Enterprise Design & Management*,   135
Advances in Intelligent Systems and Computing 261,
DOI: 10.1007/978-3-319-04313-5_12, © Springer International Publishing Switzerland 2014

The Logic of the Reference in the IT Economy

Francis Jacq

In a general context of budget control of the Information and Telecommunication Direction, in two large French companies (an airline company and a telecommunication operator), we have conducted two consulting interventions focusing on complementary areas: the first one on the existing planning of the information system developments; the second one on managing the documents of reference describing the core processes of the company.

We made the two following observations:

- In the first one, although the deliverables were of poor quality and exceeded their deadlines, the budgets of these projects are systematically reviewed upward the following years;
- In the second one, arbitrary decisions to reduce the cost of maintaining documents of reference and rigid management of access rights caused major negative impacts on business: ignorance or loss of documents and non-compliance with contracts.

We observed the references making the decision were not directly related to the business objectives. They were linked with management issues: reducing uncertainty or risk for the decision-maker, protect the hierarchical authority. Considering these two different situations:

- In case of uncertain ultimate success of a project, the decision-maker may decide to finance the acquisition of a "reference" - ISO norm, software used by all competitors - providing them with an institutional guarantee of success. They also keep funding it the following years. This institutional guarantee is considered as a powerful magical instrument ensuring success. The symbolic emotion created by these magic potions hides the true fundamental business issues to decision-makers.

Francis Jacq
Semiologist
France

P.-J. Benghozi et al. (eds.), *Digital Enterprise Design & Management*,
Advances in Intelligent Systems and Computing 261,
DOI: 10.1007/978-3-319-04313-5_13, © Springer International Publishing Switzerland 2014

- In case a business process is already working, the documents of reference lead the best practices and give authority to experts. To protect their power –their hierarchical power as only reference, managers implement restrictive access rights, witch limit linking between documents. So the managers alter the performance of their business.

In both cases, harmful defensive behaviors of managers spoiled the use of reference as a sharing guarantee witch lead the collective action. We solved these different issues using the same methodology: to overcome the references used by the management to make a decision and to rebuild new references. Our interventions have been conducted in two phases:

1) With groups including both managers and employees, refocus on the main business issues: business objectives of a project, needs of specific documents of reference.
2) Identify useful "references" for these main issues, and gradually build a consistent "set of references".

- In case of uncertain ultimate success of a project, the "set of references" give to the decision-maker the specific financial metrics to allocate a budget along one or several years: "The Project Economic File"
- In case of best practices lost or ignorance, the "set of references" organize the process which give to the documents a well-known reference guarantee by addition: expertise n°1, expertise n°2, validation n°1, validation n°2, manager n°1, manager n°2. The "set of references" indicate also the distribution area of the documents of reference and the links between them.

Conceptual Design and Simulation of an Automotive Body Shop Assembly Line

Remiel Feno, Aline Cauvin, and Alain Ferrarini

Keywords: Conceptual design, assembly line, modeling, simulation, automotive body shop.

In most industrial organization, simulation based analysis is recognized as an essential tool for designing manufacturing systems. Improving the assembly line design process and analyzing performance at early stage of a project is still a major challenge. The objective is to support decision process on strategical choices regarding the future system configuration. Indeed, early design decisions typically lack formal preliminary analysis for making good decisions based on incomplete or unreliable information.

In this context, simulation and digital factory concept represents tools and related methodologies to support manufacturing design process, taking into account the whole system lifecycle (Kühn, 2006). However, these model based approaches usually focus on detailed design and less on conceptual design where key decisions are made based on few information (Robinson, 2012). This research work deals with several assembly line architectures assessment which can be declined under various configurations.

An original approach combining Business process modeling and simulation is proposed to analyze several scenarios of process layout and structures. Indeed, for each simulation project conceptual models of an assembly line are first defined to support designer's choice and solution analysis. Many practitioners have used hierarchical, modular or functional modeling concepts in conceptual design to handle the complexity of the system (Schuh and Brussel, 2003). Such best practices are however not well applied in practical and there's a need to reduce the gap between industrial practices and research. Therefore, most of actual

Remiel Feno · Aline Cauvin · Alain Ferrarini
LSIS - Laboratoire des Sciences de l'Information et des Systèmes
Université Aix-Marseille, France
e-mail: {remiel.feno,aline.cauvin,alain.ferrarini}@lsis.org

P.-J. Benghozi et al. (eds.), *Digital Enterprise Design & Management*,
Advances in Intelligent Systems and Computing 261,
DOI: 10.1007/978-3-319-04313-5_14, © Springer International Publishing Switzerland 2014

simulation tools are technical solution modeling oriented and dismiss functional aspects that can be implemented by several possible solutions. The limitations of the identified approaches are shown to be determined by the capabilities of the simulation modeling tools (Pierreval and Paris, 2001).

Our methodological approach identifies five steps involving (1) system description, (2) preliminary analysis, (3) conceptual modeling, (4) detailed design, and (5) simulation. On the one hand, three pull control assembly line representing batch and single part flows are modeled using business process modeling notation. These models are used to decide the adequate architecture to fulfill the demand. A batch flow control has been identified to fit the mixed model demand. On the other hand, a U and L shaped assembly line are described as possible configurations corresponding to the chosen assembly line architecture. The last step consists in running simulation analysis for each configuration and performing multi-criteria analysis to compare each alternative based on cost, delay, quality and flexibility. This approach enables us to formalize the design process of assembly systems from conceptual to detailed design.

Using Models for Building Strong Organizations

Bas van Gils

When Darwin wrote his opus magnum *On the Origin of Species* he most likely did not have businesses and other enterprises in mind. Yet, the adage *"survival of the fittest"* definitely applies to many modern day enterprises. At the moment of writing, it appears we are just heading out of a long economic/ financial crisis. Many corporations have folded, others have struggled to stay afloat. This poses both a challenge and a threat for most of us: we're all in this together, some will survive and others will not. A few things seems clear: organizations that want to survive must (a) adapt to changing circumstances, and (b) do so effectively (that is: fast, and continuously). We also know that these changes will touch upon all aspects of the organization: people, processes, information/ data, infrastructure, and information systems alike.

To face these challenges, organizations should adopt a holistic approach to challenges and the associated changes. This means linking strategy to execution, business to IT, and an engineering approach to change to organizational learning. Based on an extensive body of research (ranging from change management to enterprise engineering, business transformation, and enterprise architecture), as well as many organizations world-wide (in government as well as industry settings), we believe that models play a crucial role in making that happen.

In our view, models…

- …can be made at various levels of abstraction: strategy, architecture, design, …
- …should be linked across abstraction levels
- …provide a solid basis for gaining a deeper understanding of one's business and the (impact of) of challenges and change
- ...provide an indispensable tool for migration planning, risk management, and communication with all stakeholders in the face of business challenges and change.

Bas van Gils
BiZZdesign, the Netherlands
e-mail: b.vangils@bizzdesign.nl

P.-J. Benghozi et al. (eds.), *Digital Enterprise Design & Management*,
Advances in Intelligent Systems and Computing 261,
DOI: 10.1007/978-3-319-04313-5_15, © Springer International Publishing Switzerland 2014

Having an effective enterprise modeling capability along is not sufficient however. As indicated previously, an approach to change that combines engineering aspects (implementing a vision) and an organizational change/learning approach (interventions and supporting people to act effectively towards achieving organizational goals) are essential.

We will present an approach that combines methods and tools from the realm of enterprise architecture (such as TOGAF and ArchiMate), other modeling approaches (business model canvas, the decision model, BPMN, ERD) to show how a model-based approach will help you build a strong organization in the face of the business challenges that we face today.

Quantifying Risk of Acquisition Portfolios

Hassan J. Bukhari, Ricardo Valerdi, and Daniel Ward

A risk management methodology for acquisition portfolios of programs based on cost data using a conceptual process similar to the one used in analyzing mutual funds' performance is presented. Different portfolio management methodologies are surveyed and a measure to characterize risk at the portfolio level is proposed. To illustrate the application of this methodology a case study is presented that highlights the possible benefits as well as its shortcomings. The results strongly point towards the benefits associated with portfolio level planning and indicate that risks associated with acquisition can be managed by the proposed method. We also discuss the implications of portfolio risk assessment in the context of acquisition initiatives in the Department of Defense.

More specifically, following hypotheses are proposed:

- The risk index is a valid way to quantify portfolio level risk of large acquisition programs.
- The risk index can be used for relative comparison of risks between programs.
- It is possible to develop a measurement system or framework that will assess risk in portfolios with the goal to redirect resources to the more viable programs and to make critical decisions on the continuation or discontinuation of other programs.

In this paper we also review the different methods that exist to characterize risk at the portfolio level both within the Department of Defense framework and outside, compare them and then propose a new method for portfolio risk management scheme which fills a gap that exists in the current portfolio management techniques for efficiently measuring and prioritizing risks at the portfolio level.

Hassan J. Bukhari
Massachusetts Institute of Technology, USA

Ricardo Valerdi
University of Arizona, USA

Daniel Ward
US Air Force, USA

P.-J. Benghozi et al. (eds.), *Digital Enterprise Design & Management*,
Advances in Intelligent Systems and Computing 261,
DOI: 10.1007/978-3-319-04313-5_16, © Springer International Publishing Switzerland 2014

Lessons Learned from Context-Aware Services with Real Users for Real Applications in Real Spaces

Ichiro Satoh

Context-aware services are one of the most typical services in the future digital world, which are seamlessly connecting between the real and digital worlds. There have been many attempts to provide such services. However, context-aware services in real spaces have a variety of problems, which have not been discussed in academic researches. This abstract discusses problems learned from our experiences on context-aware services in real spaces and proposes solutions to them. Most of the problems are common among digital services in real or cyber spaces.

Most users could not know where they received context-aware services, or they wandered the exhibition room and happened to enter certain places and receive the services. While they were not at the places, they could not know where and what services are provided for them, because the services were available in certain context, e.g., particular locations and users. We believe that these are an essential problem of context-aware services. However, as long as we know, no researchers discussed the problem. Context-aware services, e.g., location-aware, time-aware, and user-aware, are provided only while the current contexts, e.g., locations, times, and users, are satisfy the conditions of the services. Users, who are not in the contexts of services, cannot know the services. We put visual marks on the floor in front of the places that location-aware services were.

We furthermore have many lessons learned from our experiences:

- Mobile terminals as obstacles in real spaces.
- Support to multiple users
- Supports to legacy spaces
- Rapid installation and deployment
- Notification about sensing errors

Ichiro Satoh
National Institute of Informatics, Japan

P.-J. Benghozi et al. (eds.), *Digital Enterprise Design & Management*,
Advances in Intelligent Systems and Computing 261,
DOI: 10.1007/978-3-319-04313-5_17, © Springer International Publishing Switzerland 2014

Context-aware services are one of the most typical services in the future digital world, which are seamlessly connecting between the real and digital worlds. Although there have been many academic projects on context-aware services, many potential problems in the real world have been not well known. This paper described several the problems and our solutions to them. We believe these problems are common in future digital services.

Reference

[1] Satoh, I.: Digital Value Chains for Carbon Emission Credits. In: Benghozi, P.-J., Fiammante, M., Krob, D., Rowe, F. (eds.) Digital Enterprise Design & Management 2013. AISC, vol. 205, pp. 99–110. Springer, Heidelberg (2013)

Five Key Priorities for Enterprise Architects Involved in Cloud and Data Center Projects

Lionel Mazurié

Although Enterprise Architecture is now quite well known and widely integrated into the corporate systems of large companies, certain tasks are required and should be optimized in order to successfully achieve the migration to the cloud and data centers. After going through the various architecture layers so as to avoid any tunnel vision during these critical IT changes, five key points are identified as priorities for Enterprise Architects.

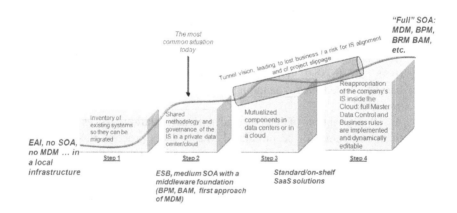

A new generation of tasks to identify and optimize services can then be launched in order to implement the usual layers of Enterprise Architecture (business, functional, applicative, and technical) according to established well-known disciplines (BPM, BRM, MDM, etc.):

Lionel Mazurié
METHOPE SI, France

P.-J. Benghozi et al. (eds.), *Digital Enterprise Design & Management*,
Advances in Intelligent Systems and Computing 261,
DOI: 10.1007/978-3-319-04313-5_18, © Springer International Publishing Switzerland 2014

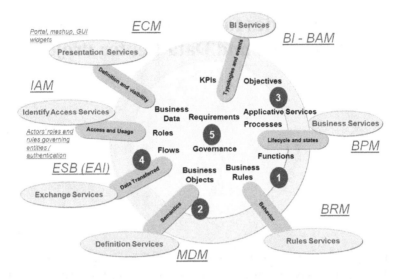

The five key priorities can be summarized as follows:

1. Modeling that takes both a static and dynamic functional view (including orchestrated and choreographed functions with associated business rules), as opposed to the use of classical business process modeling.
2. A pure Master Data approach with a company ontology incorporating a good level of granularity in order to track the data lifecycle through creation, use, business intelligence and end-of-life.
3. An efficient application inventory (As-Is To-Be), including everything from applicative projects/applications to shared and specific services (both off-the-shelf and legacy systems) associated with functions hereafter identified, including sustainable requirements to monitor specifics.
4. A full description of internal and external dataflows so as to optimize them according to the initialized Master Data process: this becomes a bottleneck in a cloud or in a data center project, representing a huge part of the IS budget that could be reduced.
5. IT Governance and Requirements Management (can be included in ITIL / CMMI global approaches) that rely on sustainable functional and technical standards to be adopted by all internal and external actors for reasons of cost, security, asset management, software licensing compliance, and homogeneity of systems in order to make changes easy.

Towards an Integrated Management of Business Models and Enterprise Architecture – Potentials and Conceptual Model

Jurate Petrikina and Karsten Zimmermann

An enterprise has to take several impediments to become the new dominant type of enterprises: the digital enterprise. In order to survive in the global competition companies need to digitize their products and services which often results in changing existing and implementing new business models (BM). As information technologies (IT) state an important trigger for new BM, a view on the structure of an enterprise and especially its IT landscape has become essential for an effective business model management (BMM). This requirement can be effectively supported by existing approaches of enterprise architecture (EA) and enterprise architecture management (EAM) as they provide a holistic and transparent view on the architecture of an organization. Therefore we conducted a study on the BMM and EAM interdependencies in order to provide potentials, requirements and a conceptual model for an integrated management of BM and EA.

Based on the systematic analysis of the existing literature on BM and EA we conclude that an integrated management of BM and EA enables a set of potentials for each of these approaches. For BMM linking BM and EA increases the transparency of relations between business and IT, leads to improved planning of resources and projects and enables the reuse of existing resources and "bottom-up" generation of new BM. The integration of BM and EA also creates transparency about the strategic direction of business development and complementing IT investments, supports the business motivation inside the IT and leads to a more adequate development of the to-be-architecture and a more detailed view on the business architecture layer.

Jurate Petrikina · Karsten Zimmermann
University of Hamburg, Germany
e-mail: petrikina@informatik.uni-hamburg.de,
    karstenzimmermann@gmx.de

P.-J. Benghozi et al. (eds.), *Digital Enterprise Design & Management*,
Advances in Intelligent Systems and Computing 261,
DOI: 10.1007/978-3-319-04313-5_19, © Springer International Publishing Switzerland 2014

During the conducted analysis we also identified that these concepts have several connection points that provide a perfect basis for their integrated management. First of all, it is a component view, which these concepts have in common as BM and EA are often described as a set of components and relationships between them. Furthermore there is a set of activities which is associated with BMM and EAM and a set of roles responsible for implementing these activities in organizations. Regarding the conceptual similarities we conclude that the integration has to be done on three levels: the component level to ensure the consistency of models, the process level to produce and deliver artifacts at the right time and the organizational level to make use of the information in decision-making. As existing approaches for the integration of BM and EA mostly cover only the component level we designed a conceptual model that also includes the process and organizational views and are planning to apply its action research implementation within a network of companies to prove it for the proposed potentials.

Why the Online Customer Reviews Are Inconsistent? Textual Review vs. Scoring Review

Wooseok Jang, Jieun Kim, and Yongtae Park[*]

Abstract. As the number of digital customers in online market grows, online customer review has become one of the indispensable information sources for customer feedback. However, the credibility of online customer review is often problematic due to the inconsistency between textual evaluation (review content) and scoring evaluation (review rating). Then, the question arises: why do such discrepancies in customer reviews arise? The primary purpose of this study is to explore the source of discrepancies between quantitative and qualitative aspects of customer review. To that end, this study hypothesizes that the discrepancy is caused by two kinds of uncertainties: reference uncertainty and heterogeneity uncertainty. First, reference uncertainty occurs if reviewers are influenced by previous reviews, especially positive ones. Second, heterogeneity uncertainty occurs when reviewers have divergent backgrounds and experiences. In order to test the plausibility of hypothesis, the discrepancy between textual review contents and rating scores is comparably measured and analyzed. In doing that, since review contents are unstructured text data, the sentiment analysis is applied to extract sentiment score from textual data. By taking the hotel service as an exemplary case, the process of theoretical and empirical analysis is elucidated.

Keywords: online customer review, textual review, scoring review, discrepancy, sentiment analysis.

Wooseok Jang · Jieun Kim · Yongtae Park
Department of Industrial Engineering, Seoul National University,
1 Gwanak-ro, Gwanak-gu, Seoul, 151-744, Republic of Korea
e-mail: {woosuk0219,hsns1234,parkyt1}@snu.ac.kr

[*] Corresponding author.

P.-J. Benghozi et al. (eds.), *Digital Enterprise Design & Management*,
Advances in Intelligent Systems and Computing 261,
DOI: 10.1007/978-3-319-04313-5_20, © Springer International Publishing Switzerland 2014

Author Index